"Dr Singh's and Prof. Parmar's new book on *Resilience and Southern Urbanism* is extremely relevant in today's times and has brought together city-specific research anchored to the rapidly urbanizing medium historic cities of India. This book certainly contributes to the growing scholarship on urban studies that are anchored in the Global South and attendant growing urbanisms discourses. The authors have created a very useful guide for urban scholars and stakeholders on the ground working for creating better urban futures. This interesting volume seeks to serve as a starting point for a new epistemological orientation on urban resilience, new debates and responses."

Dr Tigran Haas, Professor, Director of the Centre for Future of Places, KTH – Royal Institute of Technology, Stockholm; Editor of *In the Post-Urban World*, 2018

"Congratulations on your book project, *Resilience and Southern Urbanism: Towards a New Paradigm*. This book addresses an important gap in the resilience literature as it pertains to the Global South, and its important lessons for urbanism around the world. It presents significant research on the rapidly urbanizing historic cities of India, and the impacts, opportunities and dangers of this urbanisation. As such, it makes a contribution to the growing scholarship on urban studies that are anchored in the Global South. I believe it will be a very useful guide for scholars and stakeholders working to develop better urban futures."

Michael W. Mehaffy, PhD, Senior Researcher, Ax:son Johnson Foundation, Centre for the Future of Places, KTH University, Stockholm

"Historic cities of India have survived earthquakes, floods, drought, wars and pandemics carrying significant wisdoms on how to meet planetary shocks. Tapping traditional experience to be merged with global postmodern technologies would contribute to new thinking and practice on resilient urban transformation. This is the book's main achievement. It adds to a growing scholarship anchored in the South with wider global relevance. It is recommended as a guide for scholars and stakeholders working for better urban futures."

Erik Berg, BA, MA, MSc (t), Chairman, Habitat Norway

"Far from homogeneous rapid growth in Indian cities manifest in a wide array of challenges that range from deteriorating urban cores to contradicting realities in historic towns and infrastructure challenges. In response, this book offers a series of vital discussions, characterised by a new independent openness, on the contemporary condition of urbanism in India as a critical case that resonates with rapidly growing urban contexts in South Asia and the wider Global South. Through the exploitation of various perspectives and substantial theoretical underpinnings, Binti Singh and Manoj Parmar instigate an important discourse on urban resilience and draw our attention to the complexity of the urban condition in Indian cities, their challenges, and set the stage for capturing the opportunities these challenges create. This is a serious and important contribution to the global debate on urban resilience in rapidly growing contexts and is a must read for academics, decision makers, and future planning professionals."

Ashraf M. Salama, Professor of Architecture,
University of Strathclyde, Glasgow, United Kingdom

"*Resilience and Southern Urbanism: Towards a New Paradigm* by Dr Binti Singh and Prof. Manoj Parmar promises to be an important and extremely timely contribution to the discussion on urban resilience in the Global South.

The Editors correctly highlight the urgent need for a new understanding of this complex concept in contexts other than the Northern cities; they build on their extensive research and understanding of recent advancements on the subject and, through a detailed knowledge of the Indian context, offer localised interpretations of how cultural, environmental, human and economic characters might help develop more contextually significant interpretations of resilience. The Introduction sets the scene and justification for this transition, identifying significant and unique challenges to a context like no other. The fact that Indian cities are a kaleidoscope of conditions is central in the book, and fundamental to the discussion, as the authors suggest that no generalised view on resilience (a 'singular knowledge system') could be meaningfully applied. Rather, they suggest that the paradigm of 'uncertainty' is more useful. From here, the Editors identify typologies of resilience to which they dedicate extensive discussion.

The appeal of the book lies in a good balance between a strong and important idea, and a set of very detailed examples on how it can be studied and interpreted. It would be impossible to cover India's cities in one publication. This book promises a strong interpretative framework, and a few examples on how to use it in detail. The rest is up to the readers and those involved with the city, in India or anywhere else in the Global South. This book will be an important contribution to the key discussion on the future of our cities."

Dr Ombretta Romice, University of Strathclyde,
Department of Architecture, Glasgow, UK

"The shift from a predominantly rural to a predominantly urban society continues to unfold unabated. While cities have proven their worth as key arenas for social and economic development, such outcomes are never a certainty. With urbanisation and the rise of cities occurring at breakneck speeds in many countries throughout the Global South, significant effort is required to ensure that they scale accordingly. This means that they will require investment in the necessary housing, infrastructure, amenities and basic services to keep pace with their growing populations. Efforts that fall short of this will result in over-congested and overwhelmed cities. This makes rapid urbanisation and the unprecedented rise of cities one of the most important megatrends of our time. While there is still a lot that remains unknown about the causes, dimensions and challenges of urbanisation in the 21st century, this thought-provoking book examines rapidly urbanizing medium-sized historic cities in India to identify lessons from the past that can inform the cities of the future. The authors do so through the lenses of heritage and resilience. Despite these cities undergoing incredible transformations and changes, many of the traditional urban systems continue to remain intact, doing their part to alleviate the pressures of rapid urban growth. Not only does culture heritage build identity into the fabric of cities, but the authors demonstrate that it also plays a pivotal role in performing important urban services. For example, this book explores historical water systems, examining the role they played in creating a functioning urban environment and from a resilience standpoint how they have withstood the test of time. In doing so, important lessons can be learned for managing the challenges of the future (i.e. climate change). The book concludes with some suggestions on how local communities can be engaged in the planning process to build more resilient cities. In doing so, this book contributes to the growing scholarship in urban studies focused on the Global South and their attendant urbanisms. This book serves as a useful guide for scholars and stakeholders working to better understand the challenges that rapidly growing cities are facing and what can be done to better position these cities for the future. Those who read it will emerge with a renewed understanding of India's history and the important role it will play in shaping its future. From history, to policy to innovative solutions for building more liveable cities, this book has it all."

Dr Kyle Farrell, Urban Economist, Economic Pulse Analytics, Vancouver, Canada

"This edited collection stands on the shoulders of scholars who've called for a Southern urban theory, and makes an attempt to articulate the resilience discourse in Indian cities. This is a timely contribution considering that the current discourse on resilience heavily draws from the experience of the cities of the North, assuming a singular authoritative knowledge system. Situating resilience makes special sense during the COVID and post-COVID times of uncertainty when densely populated and resource-poor cities of the Global South struggle for their livelihoods. I, personally, find it intriguing that Singh and Parmar (the editors) arrive at a typology of resilience centered around culture. This calls for further investigation of resilience stories and struggles in the Southern cities, with special attention to cultural practices."

Dr Tooran Alizadeh, Associate Professor of Urbanism,
The University of Sydney, Sydney, Australia

"In *Resilience and Southern Urbanism*, Dr Binti Singh, Prof. Manoj Parmar, and their respective collaborators show us that we can no longer glibly discuss the trials and tribulations of evolving cities without reference to the Global South and the scholarship of the world's second-most-populous nation. Their rich melding of cultural, environmental, economic, and human-scale factors is this interdisciplinary paradigm we need, honouring important issues of the day without resorting to trendy sameness. The resilience theme is equally compelling in these times, but most important is this book's stage: a subcontinent where the authors' granular and critical examination provides an approach deserving of respect around the world."

Charles R. Wolfe, Founder and Principal, Seeing Better Cities Group, Newbury, UK; Affiliate Guest Scholar, Centre for the Future of Places, KTH Royal Institute of Technology, Stockholm Affiliate Associate Professor, University of Washington, Seattle; Author, *Sustaining a City's Culture and Character* (2021), *Seeing the Better City* (2017), and *Urbanism Without Effort* (2013, rev. ed. 2019)

RESILIENCE AND SOUTHERN URBANISM

This volume studies the urbanisation trends of medium-sized cities of India to develop a typology of urban resilience. It looks at historic second-tier cities like Nashik, Bhopal, Kolkata and Agra, which are laboratories of smart experiments and are subject to technological ubiquity, with rampant deployment of smart technologies and dashboard governance.

The book examines the traditional values and systems of these cities that have proven to be resilient and studies how they can be adapted to contemporary times. It also highlights the vulnerabilities posed by current urban development models in these cities and presents best practices that could provide leads to address impending climate risks. The book also offers a unique Resilience Index that can drive change in the way cities are imagined and administered, customised to specific needs at various scales of application.

Part of the *Urban Futures* series, the volume is an important contribution to the growing scholarship of southern urbanism and will be of interest to researchers and students of urban studies, urban ecology, urban sociology, architecture, geography, urban design, anthropology, cultural studies, environment, sustainability, urban planning and climate change.

Binti Singh is an urban sociologist. She holds a PhD in urban studies and an MPhil in Planning and Development from the Indian Institute of Technology, Mumbai, India. She is Associate Professor and Dean (Research and Academic Development) at KRVIA, Mumbai, and is engaged in diverse international research programmes. She is also Book Series Editor of the series *Urban Futures* and Associate Editor of *Oxford Urbanists*. She has contributed to several peer-reviewed research journals and edited books.

Manoj Parmar is currently Director, KRVIA, Mumbai. He holds a Bachelor's degree in architecture from the L.S.R. School of Architecture, India. He also holds a M. Arch from the University of Miami, Florida, USA. He has been teaching at KRVIA, Mumbai, since 1992. His academic interests include theoretical writing on architecture and urbanism.

URBAN FUTURES

Series Editor: Binti Singh,

Associate Professor and Dean (Research and Academic Development), KRVIA, Mumbai, India

Uncertainty is the "new normal" for cities of the future. At this juncture our cities (both in the developed global North and in the developing and rapidly urbanizing global South) are at the crossroads of unprecedented challenges and cautious choices. Cities worldwide are also bound by the general guidelines of the Sustainable Development Goals guiding our collective urban futures. No matter how much we think of our urban futures as a "collective", questions of structural inequalities in economic opportunities, access to basic services, escalating vulnerabilities to climate change risks and justice compel us to look for grassroots, local responses and solutions. Somewhere, cities seemed to have lost the plot while traversing their soft and hard boundaries. The *Urban Futures* series attempts to draw the contours of our collective urban futures, identifies and articulates the pressing challenges cities face globally. It provides a starting point to some of the responses that we could adopt for a better and a more inclusive future.

The series covers a wide range of interdisciplinary themes including urban design, urban digitization, governance and data city, technology, social inclusion, social infrastructure and public health, city life and public spaces, consumption and culture, urban planning, mobility, environment, public infrastructure, sustainable development, localization of SDGs, women and urban development, urban economy and work, and peri-urban development.

EMERGING WORK TRENDS IN URBAN INDIA
COVID-19 and Beyond
Nidhi Tandon, Pratyusha Basu, Omkumar Krishnan and Bhavani R.V.

RESILIENCE AND SOUTHERN URBANISM
Towards a New Paradigm
Edited by Binti Singh and Manoj Parmar

For more information about this series, please visit: www.routledge.com/Urban-Futures/book-series/UBF

RESILIENCE AND SOUTHERN URBANISM

Towards a New Paradigm

Edited by
Binti Singh and Manoj Parmar

Routledge
Taylor & Francis Group
LONDON AND NEW YORK

First published 2022
by Routledge
4 Park Square, Milton Park, Abingdon, Oxon OX14 4RN

and by Routledge
605 Third Avenue, New York, NY 10158

Routledge is an imprint of the Taylor & Francis Group, an informa business

British Library Cataloguing-in-Publication Data
A catalogue record for this book is available from the British Library

Library of Congress Cataloging-in-Publication Data
A catalog record for this book has been requested

ISBN: 978-0-367-56352-3 (hbk)
ISBN: 978-1-032-23014-6 (pbk)
ISBN: 978-1-003-09846-1 (ebk)

DOI: 10.4324/9781003098461

Typeset in Times New Roman
by Apex CoVantage, LLC

This book is dedicated to all stakeholders who are working relentlessly both individually and collectively to build better urban futures.

CONTENTS

ILLUSTRATIONS

Figures

Boxes

Tables

AUTHOR BIOGRAPHIES

Editors

Binti Singh is an urban sociologist. She holds a PhD in urban studies and an MPhil in planning and development from the Indian Institute of Technology, Mumbai, India. She is Associate Professor and Dean (Research and Academic Development) at KRVIA, Mumbai, and is engaged in diverse international research programmes. She is also Book Series Editor of the series *Urban Futures* and Associate Editor of *Oxford Urbanists*. She has contributed to several peer-reviewed research journals and edited books. Her articles have also featured in *Domus India* and *Business World Smart Cities*. Her research engages with questions on the built environment, urban policy and governance, urban trends and urban theory. She has authored four books including *Culture, Place, Branding and Activism* (2018), an ethnographic study on Lucknow, the capital of Uttar Pradesh in India, and *The Divided City Ideological and Policy Contestations in Contemporary Urban India* (2018). Her third book *Smart City in India: Laboratory, Paradigm or Trajectory?* (co-authored with Manoj Parmar) was published in November 2019 and was widely discussed including at the University of Strathclyde, Glasgow, UK, ITC University of Twente, the Netherlands, the *Alliance Francaise* Mumbai and the *Kala Ghoda* festival, 2020 in Mumbai. Her fourth book *How Will India Fix Her Urban Future?* (co-authored with Sameer Unhale, Urban Practitioner and currently State Joint Director, Directorate of Municipal Administration, Government of Maharashtra, India) was published in 2020.

Manoj Parmar is currently Director, KRVIA, Mumbai. He holds a Bachelor's degree in architecture from the L.S.R. School of Architecture, India. He also holds an M. Arch degree from the University of Miami, Florida, USA. He has been teaching at KRVIA, Mumbai since 1992. His academic interests include theoretical writings on architecture and urbanism. He has also been in private practice of architecture, urban design since 1992. He has worked on numerous private and public housing/institutional commissions across the country. He has been actively involved in redevelopment projects across the city of Mumbai. He has also executed projects in Dubai, Malaysia and Singapore. He has been a part of drawing up development guidelines for slum redevelopment in the northern suburbs of Mumbai. He has presented his work at the young architect's forum at UDRI & CEPT University, India. He currently teaches Architectural Theory and Thesis at the undergraduate level and Housing Studio and Thesis at the postgraduate level. He has contributed several academic papers to peer-reviewed journals, magazines and edited books. He has also presented his work in various international conferences. He is the co-author of *Smart City in India: Laboratory, Paradigm or Trajectory?* (2019, co-authored with Binti Singh).

Contributors

Pradipta Banerjee, is Director, Centre for Urban Sciences and Engineering and Professor of Civil Engineering, IIT Bombay, Mumbai, India. He is known internationally for his developments on Bridge Asset Management using structural health monitoring, especially for the Indian Railways and Network Rail in the UK, and vibration control of buildings for earthquake ground motions using innovative strategies. He has also developed technology that has been and is being implemented in India and abroad in the areas of disaster resilience and asset management. His commitment to capacity building in applied science, engineering and technology in developing countries has seen him developing relationships between the IITs and Academic Institutions in Nepal, Bhutan, Bangladesh, Sri Lanka, Cambodia and Vietnam to foster capacity building in ASET.

Jamshid Bhiwandiwalla is Associate Professor at KRVIA, Mumbai, India. With a Master's in architecture as well as in ancient Indian culture, he has been a recipient of the MASA Best Teacher Award for the subject of Humanities in the year 2010, thereby formally guiding students since the last 20 years. Equally old is his independent architectural practice associated with Heritage Conservation, largely restoring community institutions like Fire Temples and Convalescent Homes. His major area of study is the Conservation and Management Plan for the Heritage Pilgrim town of Udvada, Gujarat, where he was officially appointed as the consulting architect for the Gujarat Govt. Presently, he is Member Appointed on the Heritage Conservation Cell of the Udvada Area Development Authority.

Further, he has curated interactive exhibits in Heritage structures such as the Zoroastrian Information Centre, Udvada again for the Gujarat Govt. as well as converting the birthplaces of Sir Jamshetji Jejeebhoy, Sir Jamshedji Tata and Dadabhai Naoroji into Memorial Museums in Navsari, Gujarat. The chapter on Water Resilience is part of such an involvement of a studio associated with the water systems of historic cities in India. Presently, he is a partner with the yearlong Fellowship Research Program on Urban Water Resilience funded by the Kamla Raheja Institute of Architecture and Environmental Studies.

Sandeep B. Menon is an academician and landscape architect. He is a core faculty member at KRVIA, Mumbai, India. He holds a master's in landscape architecture and a bachelor's in architecture from the School of Planning and Architecture, New Delhi, India. He was the Design Director at Integral Designs International Studio Pvt Ltd, New Delhi. He has worked on landscape design and master planning projects of varying scales in various bio-geographies. His work often addresses the ideas of natural processes, local ecology and native planting. His interests range from ecological urbanism, landscape ecology, sustainable urban water management, ecological corridors and wetland systems.

Vikram Pawar is Professor at KRVIA, Mumbai, India. He has conducted design studios and guided postgraduate as well as graduate students. He has been teaching building technologies and conservation sciences at KRVIA and has been instrumental in restructuring and shaping the M. Arch course in Architecture and Urban Conservation at the Mumbai University. With over 20 years of professional experience, he is a founding partner-director at Studio Architecture Heritage Environment Consultancy (SAHEC). He is also a co-founder of Water Environs (WE), a collaborative of SAHEC and Educated Environments (EdEn) working towards revival of Mumbai rivers since 2010. He has coordinated and curated *Housing and Architecture*, a reference compendium for MHADA – YASHADA programme for MHADA officials and staff and authored an essay on 'Housing for Poor' for InHAF's programme on Bamboo for Housing Poor and has also authored a document – 'Participatory In-situ Slum Rehab – two case studies (in Pune and Nagpur)'.

FOREWORD

Amidst Strives to Adapt to Climate Change's Impacts, the Urban Poor's Needs Are Often Overlooked

Climate change will adversely impact the weakest and the most vulnerable sections of urban societies. Due to their location in areas of hazard risk such as steep slopes, wetlands or flood-prone areas and the widespread usage of non-durable building materials, marginalised settlements are often very vulnerable to disaster. Increased frequency of natural hazards and sea-level rise are expected to be among the most severe impacts of climate change in India. Urban poor in large cities are therefore likely to be most affected.

Paradoxically, their vulnerability to those impacts may increase the danger to be forcibly evicted as it frequently serves as a pretext for local authorities to clear 'unsafe' settlements. Past experiences with slum redevelopment schemes in India proved that private real estate developers and builders are interested in developing such sites, albeit without taking poor inhabitants' needs into account. Amidst strives to adapt to climate change's impacts and to meet climate targets, the poors' needs are therefore mostly overlooked.

However, the vision for cities of the 21st century must be inclusive! Therefore, a need arises to strongly enhance marginalised communities' resilience to prevent vicious cycles of disaster risk, poverty and vulnerability. Consequently, the most critical aspect of disaster risk management is to involve affected communities.

There are development goals of addressing climate change impacts and pursuing low-carbon development pathways to meet the objectives of the Paris Agreement and India's Nationally Determined Commitments (NDCs). Addressing climate change impacts is also of relevance under the Sustainable Development Goals (SDG) and in particular Goal 11, which aims at cities and human settlements to be inclusive, safe, resilient and sustainable.

The focus of contemporary planning and design needs to shift towards new paradigms of enabling increased resilience, quality of life and well-being, especially in poor, underserviced and often informal urban communities.

To further this programmatic shift, urban professionals need to understand relevant government schemes, develop designs with urban poor as clients, elaborate bidding documents for projects and monitor and evaluate ongoing projects and their design elements.

In consequence, there is also a need for a paradigmatic shift in the education of graduate students in spatial planning and design and a need for training of urban professionals from different backgrounds. The complex interdependencies of climate change adaptation and urban poverty have to be highlighted to urban professionals, and they need to be trained in proactively dealing with them.

In this book, based on city-specific case studies, each author has attempted to arrive at a typology of resilience. Furthermore, the editors Binti Singh and Manoj Parmar have contributed to the growing scholarship on southern urbanism, as they carry forward from their previous book 'Smart City in India Urban Laboratory – Paradigm or Trajectory?', published by Routledge in 2019.

In sum, this comprehensive work highlights important issues by centring on culture in the narrative of resilience, and it shows the way forward to build safe, resilient, inclusive and sustainable cities as envisaged in the SDGs.

Tania Berger
Coordinator of 'Building Resilient Urban Communities' project (www.breucom.eu/)

FOREWORD

The ethos of Global urbanism is under current review to ascertain its validity as a 'one size fits all' paradigm and has led to certain assertions attempting to invalidate it by proposing alternate theories based on observations of extreme variants in cities, thereby questioning the concept of 'planetary urbanisation' (a view, put forward by Brenner and Schmidt who have erased differences among cities by putting them on a common platform). The roots of the movement of this recent alternate view, 'southern urbanism' lie at the African Centre for Cities, University of Cape Town, an interdisciplinary space, dedicated to rethinking on urban things and is picking up favour as well as space by several writers and thinkers who are led into inquiring about it by understanding and questioning their own urban context. The importance of southern urbanism emphasises that theories framed outside of the earlier understood 'third world' or 'developing countries' domain do not qualify for outright adoption of plans, processes or projects that shaped the 'developed world', since these were founded on the global capitalist order and its outcomes, hence unsuitable in shaping urban sector policies and plans for the Global South, and may even be a deterrent to 'Resilience' in cities by adding to them the complexities of global urbanism.

As an example, segregated Land Use Planning (the domain of Master Plans in India) created opportunities for the automobile industry to flourish by creating distances for the population to negotiate, in similar vein as Smart Cities intend propagating and marketing devices and sensors for growth of the ICT industry in a globalised business environment. Urbanism practised and experienced in cities of the Global South (implying Africa, especially sub-Saharan Africa, Asia and Latin America) offers a different context and Western-informed urban theories do not provide the right guidance on urban development to them.

In a broad understanding of this background, the editors of the book, Dr Binti Singh and Prof. Manoj Parmar, have undertaken the colossal task of questioning the foundation of a monolithic understanding of urbanism, choosing to focus on the experiences and diversity of second-tier historic cities in India which are in line of succession to become large and display distinct local flavour for assessing if new thinking is required on their resilience, and for initiating the epistemological debate.

Effects of the COVID-19 pandemic have been a global challenge, yet local responses to it have been in sharp contrast in the developed north to those in resource-poor cities of the south indicating their inbuilt resilience. A demonstrated ability to respond to or recover from economic, environmental, social and institutional challenges to sustain human well-being or inclusive growth over long periods of time is indicative of resilience in the cities of the Global South, so is assimilation of different cultures, non-agitative, local and profound responses to everyday struggles of livelihood, basic services, inadequate housing, a host of other discomforting factors which do not promote aggressive behaviour in southern urbanism despite significant variation in geographic location, social fabric, culture, heritage and economy; the decimating power of the 'political economy' cannot override contexts on which the urban process unravels.

At this juncture, I am tempted to note that an American author, Dave Prager, in his book 'Delirious Delhi' gives in first-person account of his and his wife's everyday experiences of numerous instances where they would get flustered, yet marvelled at the citizens' spirit of patience and acceptance of situations not under their control, indicating perceptions of people from two different worlds and their starkly opposed reactions and responses, indicating differences in their resilience.

In a way, the logic of neoliberal capitalism pervading urban space in cities of the Global South is a point to be disqualified.

Of the book's six chapters, the first two constitute the foundational questions on which the epistemological debate rests, focusing on existing values and resilient systems prevalent in the medium historic towns, whether they signify resilience through their cultural heritage, water or a coastal city ecology, and whether these dimensions help position the debate on adoption of the alternate 'Southern Urbanism' theory; to assess if the imposed development models (e.g. Smart Cities) add vulnerabilities to resilience, as also the need for rejection/redesign of those strategies which endorse planetary urbanisation through remote access thereby depriving cities of the citizen-connect/community building; technosocially whether such ingress hinders resilience, a rethinking would be necessary. Similarly, the medium historic living cities transiting into modern economy, with an ushering in of western models and digital control, is likely to raise questions on the dismissal of sociocultural contexts of historic cities accompanying the discard of sympathetic design and cultural practices; in short, will modernity trample tradition which remained for centuries and prompted resilience and gave important lessons on what gives a city the impetus for sustainable growth and well-being, or, what makes it lose its resilience?

Subsequent chapters open out the variation in typology of cultural, water and ecological resilience, respectively. The first case study on Nashik is indicative of the avalanche of projects imposed on the city for the growth of its economy, such as, expressways/SEZs/Entertainment Zones, and questions the status of retention of the tangible and intangible cultures, thereby the status of its cultural resilience.

For water resilience, Bhopal is under the scanner for its self-sufficiency in that domain and if any likely overburden may impede resilience; similarly for a resilient coastal ecology, Kochi has been taken up to ascertain if the Vembanad estuary on which the port is in active performance would hinder or become non-resilient to the biotic and abiotic interactions traced over centuries.

The authors have taken on a highly challenging subject, and I fully endorse the lurking trends overseen for a unitary and compulsive way of assessing and ascertaining urban space and urbanism which to my mind is an imposition on human thinking not backed by scientific reasoning, data or observation. Their work will be a contribution to our literature on Southern Urbanism. And I am proud of the timings in which the authors have delved so deeply to explore an important urban study impacting our resilience in the near future. My hearty congratulations to them for their early entry into an arena which, if left unattended, can impact some questionable urban futures.

Professor Veena Garella
Freelance Urban & Regional Consultant
Professor and ex Head of Department – Urban Planning
School of Planning and Architecture
New Delhi, India

ACKNOWLEDGEMENTS

The idea of this volume incubated while working on our previous co-authored book *Smart City in India: Urban Laboratory, Paradigm or Trajectory?* published by Routledge in 2019. Working on the research for that volume, we had first-hand exposure to the rapidly urbanising landscape of India, especially the medium historic cities of India. Recent urban programmes like the Smart City Mission in India launched in 2015 have subjected these cities to massive and rapid transformations. An academic reflection to examine the gaps and challenges was imminent. Alongside, working on the Building Resilient Urban Communities (BReUCom) research programme supported by the European Union brought us to the point of a deeper understanding of resilience especially anchored to the experience of most vulnerable cities of the Global South. The questions of cultural practices and social structures provided the foreground for our understanding of our built environment motivated with new thinking and growing scholarship on building better urban futures. The pandemic of 2020 reaffirmed our ongoing research agenda around the questions of inclusion, culture, urban form and urban theory. Inputs from subject experts, knowledge partners, friends and colleagues from around the world greatly facilitated our academic trajectory. Notable among them are Prof. Sergio Porta and Prof. Ombretta Romice from the University of Strathclyde, Glasgow; Prof. Rahul Mehrotra, Harvard University Graduate School of Design; Prof. Pradipta Banerjee, Centre for Urban Sciences and Engineering, IIT Bombay, Mumbai; Chuck Wolfe, Tigran Haas, Michael Mehaffy and Andrew Karvonen of KTH – Royal Institute of Technology, Stockholm; and Erik Berg, Chairman, Habitat Norway. Tania Berger (Donau University, Krems, Austria) and Coordinator of BInUCom and BReUCom programmes have always been an encouraging force and together we have endeavoured to carry this research forward to cover many more cities and develop a robust scholarship on urban studies based on the southern experience. We also endeavour to achieve a wide dissemination of resilience understanding and generate awareness among multiple stakeholders, and strive to find local solutions to pressing global problems. This volume is also the result of several rounds of discussion with colleagues of the Postgraduate Program at KRVIA, Mumbai who have contributed as chapter authors.

ABBREVIATIONS

AI	Artificial intelligence
CAA	Constitutional Amendment Act
CCPI	Climate Change Performance Index
CEO	Chief executive officer
CH	Cultural heritage
COVID-19	Coronavirus Disease 2019
DMIC	Delhi–Mumbai Industrial Corridor
DP	Development Plan
ECG	Electrocardiogram
FHS	Future Health Systems
GA	The great acceleration
GDP	Gross domestic product
GHG	Greenhouse gases
GIS	Geographic information system
GOI	Government of India
GPS	Global Positioning System
ICCC	Integrated Command and Control Centers
ICH	Intangible Cultural Heritage
ICOMOS	International Council on Monuments and Sites
IGBP	International Geosphere-Biosphere Programme
IIHMR	Institute of Health Management Research, Bangalore
IISc	The Indian Institute of Science
IOT	Internet of Things
IPCC	Intergovernmental panel on climate change
IT	Information technology
JNNURM	Jawaharlal Nehru National Urban Renewal Mission
NDMA	National Disaster Management Authority
NGOs	Non-governmental organisations
NITI *Aayog*	National Institution for Transforming India
NMC	Nashik Municipal Corporation

NPCA	National Plan for Conservation of Aquatic Ecosystems
NRDWP	National Rural Drinking Water Programme
NWCP	National Wetland Conservation Programme
ODF	Open defecation free
ODK	Open Data Kit
OECD	The Organisation for Economic Co-operation and Development
RIBA	The Royal Institute of British Architects
SAHEC	Studio Architecture Heritage Environment Consultancy
SANDRP	South Asia Network on Dams, Rivers and People
SBA	*Swach Bharat Abhiyan*
SBM	*Swach Bharat* Mission
SCM	Smart City Mission
SCMS	School of Communication and Management Studies, Cochin
SDG	Sustainable Development Goals
SEZs	Special Economic Zones
SRC	Stockholm Resilience Centre
SSC	Smart and Sustainable City
TRPs	Television Rating Points
UCLG	United Cities and Local Governments
UK	United Kingdom
ULBs	Urban Local Bodies
UNESCO	United Nations Educational, Scientific and Cultural Organisation
UNITU	United Nations International Telecommunication Union
URDPFI	Urban and Regional Development Plans Formulations and Implementations
WE	Water Environs
WIPRO	Western India Palm Refined Oil Limited
WSUD	Water-sensitive urban design

1

INTRODUCTION

Binti Singh and Manoj Parmar

Resilience, as a construct, has myriad connotations. Viruses, for example, are very resilient; they are robust enough to endure strong external attacks and adaptable enough to keep to their environment. The coronavirus that has wreaked havoc in most part of 2020 not only by killing millions of people around the world but also by bringing the world to a complete halt with major disruptions in the ways human beings live their lives is a case in point. By the time a vaccine is invented and the pandemic caused by this resilient virus is controlled, a lot of damage would already have been done. When we think of ourselves as collectives, as systems or as societies the term resilience takes a rather positive connotation. We celebrate resilient societies like the Japanese society that has withstood natural disasters repeatedly or Varanasi as the oldest and most resilient living city in the history of human civilisation. Related terms like sustainability have made way into a common lexicon. At the outset, let us understand the fundamental difference between resilience and sustainability – the meanings of which are embedded in culture. Rómice et al. (2020) define resilience as descriptive rather than a normative concept: it is undirected and has no final ultimate goal. It is typically an evolutionary mechanism whose only payoff is to stay in the game. They explain that a resilient phenomenon, in fact, is not necessarily a good one. In the book 'Master Planning for Change', Rómice et al. (2020) allude to a 2014 talk at the Australian Academy of Science where Brian Walker discussed this topic in the realm of ecological planetary boundaries, on existing patterns of growth and resource depletion, that he called the water-energy-food nexus. The goal of resilience focuses on the liberty of the system to keep self-organising without breaking down its basic structure of fundamental functionalities, controls and feedback (shepherds require sheep to remain in a flock, and not scatter around randomly and get lost in the bush). Further, the goal of sustainability is an agreement on what constitutes the common good, the desirable trajectory of change is searched, achieved and continuously reviewed (shepherds want sheep to

 DOI: 10.4324/9781003098461-1

eventually reach the Alpine pasture). In addition to that, neither resilience nor sustainability alone is enough and reconciling both is a cultural problem. We need a city that is both resilient and good (for details see Rómice et al., 2020, pp. 236, 37).

This book attempts to understand resilience embedded in the particularities of the urbanisms of Indian cities. We take them as representatives of southern urbanism, chosen carefully from the vast heterogeneous and diverse cultural geographies of the country. We are aware that many more cities located South of the Hemisphere could be part of this narrative. These omissions are noted and will be covered in a sequel to this volume in our quest to develop a robust scholarship on southern urbanism centred on the questions of cultural practices and community resilience. The rationale behind this volume is that the current discourse on resilience heavily draws from experience of cities of the north. Percolated globally as a singular knowledge system, this can perhaps qualify as an offshoot of planetary urbanisation. It is no wonder therefore that prevailing global perspectives on resilience attempt to establish uniform trends, identify patterns of vulnerability, similarities in evolution, configuration and political economy. The idea of planetary urbanisation entails that 'macro-trends have propelled urban processes into the fast lane, and the territorially bounded city has been eclipsed by urbanisation whose uneven development is so entangled, scope so expansive and morphology so complex and variegated, that it is nothing other than planetary' (Brenner & Schmid, 2015). The macro-urban processes are often those theorised as following the logic of neoliberal global capitalism. For instance, critical urban theory essentialises that cities worldwide are part of the global capitalist order and their urban processes are consequent outcomes of the larger global order. Uniform and monolithic understandings of urbanity, modernity, development and now resilience dismiss the specificities of cities of the Global South. The importance of cultural practices, shared identities, symbols, everyday urbanism practised and lived by citizens daily could go a long way to determine the larger goals of sustainability, inclusion and smartness in the urban landscape.

Section 2: Resilience and Southern Urbanism: A Theoretical Framework

In the present times, our cities face many paradoxes. Firstly, cities worldwide are bound by the general guidelines of the Sustainable Development Goals (SDGs) and our collective urban futures. No matter how much we think of our urban futures as 'collective', questions of structural inequality to economic opportunities, access to basic services and justice compel us to think again.

Simone (2001) argues that urban residents themselves 'appear increasingly uncertain as to how to spatialize an assessment of their life chances – that is, where will they secure livelihood, where can they safeguard protection and be looked after, where will they acquire the critical skills and capacities?' Situating resilience in this generic of 'uncertainty' makes sense for southern cities that are still geographies of everyday struggles for livelihoods, access to basic services and security unlike their northern counterparts. The COVID-19 pandemic since early 2020 has wreaked havoc globally. Yet, the ways in which cities of the developed north have responded are starkly in contrast to the implications of the pandemic for densely populated and resource-poor cities of the Global South. In the Global South, resilience takes a different meaning based on context-specific experiences and altogether different challenges posed by its cities. Secondly, the pandemic has reopened a can of questions forever left unanswered. It has exposed deep fractures and divisions of our already segregated societies. These include compartmentalisation between mental labour and manual labour, the digital divide, class divisions and the escalating miseries of migrant informal labourers. Thirdly, density that acts as a jeopardising force when it comes to the spread of contagious diseases like the COVID-19 now or the plague, cholera, tuberculosis or Spanish flu at other points in history has rendered our densest cities as non-places, bereft of their fundamental character. Indeed, what is a city without its density! Social distancing forcefully imposed on city life by a contagious disease runs in total contradiction to the dense social lives that city life offers. In the words of Friedmann (2010), place becomes 'place' only where and when social life saturates. Everything else, by that definition, are non-places. Density works exceptionally well for optimal utilisation of scarce resources, lowering carbon footprints and contributing to climate resilience in the long run, especially true for cities in developing economies like ours. Fourthly, the major disruptions in global supply chains, exports of essential goods and mobility caused by the pandemic have driven us to rethink our growth trajectories, consumption patterns and resource dependency with a call to become more self-reliant and a move towards what many describe as (de)globalisation. It is policy and practice response at the national and city levels, working towards an urban equality agenda and a new social contract theory that will perhaps help in realising the globally acknowledged SDG 11 of building safe, inclusive, resilient and sustainable cities. Further, the Urban Crisis Charter of the Global Alliance for Urban Crises presented at the World Humanitarian Summit highlights that resilience is a core goal for the UN's New Urban Agenda (UN Habitat, 2016).

The choices that we make from now on will greatly impact the narrative of resilience. For example, the flight of people to lower density sprawl (as is becoming more evident after being hit by the pandemic) spurs up our dependence on the car. If the post-pandemic city curbs mass transit and public transportation systems and steers focus on building more car-centric suburbs, cities again will be at the crossroads of difficult choices. All efforts of creating hyper cities, 15-minute cities with proximity between places of stay and work to curtail travel time, will come to a halt. The shift of priority from public transport to the private vehicle has escalating costs on environment and social life too. It is evident that we are faced with difficult choices that in turn are dependent on our capabilities and resource availability and are deeply embedded in our social structures and cultural practices. The Japanese cultural practice of *Kintsugi* or gold slicing/mending could be seen as a physical manifestation of resilience. It essentially is a practice of art repair of broken items like vessels with golden adhesive that enhances the broken lines making the piece unique and more beautiful, accentuating its value, imperfections and breaks. Its Indian equivalent is the quilt or *Katha* stitched together with worn out and old pieces of clothing to exude warmth and an emotional connect.

In recent times, urban scholars have pointed out the inadequacies and fallacies in existing urban theory as a singular knowledge system. Studies anchored to the Global South have contributed to a robust scholarship on southern urbanism. Concepts like occupancy urbanism (Benjamin, 2008), insurgent citizenship (Holston, 2008), aesthetic governmentality (Ghertner, 2010), speculative urbanism (Goldman, 2011), subaltern urbanism (Roy, 2011b), telescopic urbanism (Amin, 2013), multiplicities of governance regimes (Schindler, 2014), the near-south (Simone, 2014), entangled urbanism (Srivastava, 2014), ordinary cities (Robinson, 2006) and worlding of cities (Roy & Ong, 2011) have helped develop what Schindler (2017) calls a veritable lexicon that identifies and describes phenomena in southern cities. In addition, the 'political economy is not an overriding context within which urban processes unfold, but rather it is always already co-constituted with the materiality of Southern cities' (Schindler, 2017). In the book Urban Theory Beyond the West, Tim Edensor (2012) raises similar theoretical concerns about cities throughout the world. Reworking urban theories that are 'Western' (or anchored in the developed north) offers useful ideas like 'De-centering the City' through consideration of the diversity and heterogeneity of city life and investigates how urbanity is differently perceived and experienced based on conceptual reflection of cities from around the world. The doom-ridden representation of cities of the Global South as sites of perpetual meltdown, decay, overcrowding and violence cries for serious revision. Cities in the Global South and attendant southern urbanism have distinguishing evolution patterns, processes and historicity. The particularities and the sheer diversity of small and medium historic cities of India carry significant local flavour. With growing global linkages, these smaller cities are in the throes of rapid transition, serving as laboratories for urban experiments that call for new thinking about resilience.

The questions of cultural practices inbuilt in the everyday urbanism of Indian cities are inextricably tied to the questions of resilience. Scholars argue that resilience of social ecological systems is its ability to absorb changes without having to lose its integrity in the system, form or function (Leichenko, 2011). Scholars like Folke et al. (2002) and Ernstson (2010) therefore explain that the ability of the system to maintain, adapt and nurture its innate strengths without losing out on the opportunity for reorganisation and memory is a test of resilience. With the example of Sahet Alqaryoun (Alqaryoun square) located on the southwestern edge of the Ottoman part of the city of Nablus in harat (or quarter), Assi (undated) explains their impeded memories, loaded symbols and codes that people can identify themselves with. Appropriated over the years, Sahet Alqaryoun as an open public space represents a public history whose value resides in the complexity of their structures, drawing meaning from the interaction between monuments, houses, meeting places and workplaces, movement patterns, social habits and ritual commemorations. Shail Mayaram's book *The Other Global City* (2009) breaks the myth of a monolithic understanding of the global order and its attendant challenges like resilience through a historical-ethnographic exploration of inter-ethnic relations in the 'other global' cities of Cairo, Beirut, Istanbul, Bukhara, Lhasa, Delhi, Singapore and Kuala Lumpur.

A number of urban studies scholars, sociologists, ecologists and anthropologists have stressed the importance of deploying a cultural ecology perspective to better understand urbanisation, urban processes and urbanism in the Indian context (Baviskar, 2007; Srinivas, 2004; D'Souza & Nagendra, 2011). Gidwani and Baviskar (2011) as well as Ostrom and Nagendra (2014), Seema Mundoli et al. (2017) and Parthasarathy (2011) have engaged with the idea of urban commons in re-theorising cities in India from an ecological perspective. Mukherjee's volume on *Sustainable Urbanization in India* (2017) draws from several Indian case studies to bring in political, cultural and geographic understanding of practices, place-making and environmental degradation that is now a characteristic feature of Indian urban development. Based on an empirical study of two lakes – *Kaikondrahalli* and *Saul Kere*–situated in the peri-urban interface of Bangalore city, Sen and Nagendra (2018) argue that the application of a range of meanings and sociocultural practices coupled with a sense of belongingness transform built environments in southern cities into shared social spaces and sustainable urban commons. The nature of connection that these people have with the lakes is contrasting yet fascinating to the extent that it allows us to rethink the given versions of third-world environmentalism. Since historical times, the questions of inclusion, heritage, place-making, traditional livelihoods and cultural associations of communities are embedded with ecologies like rivers, lakes and forests (for details, see Nagendra, 2016; Unnikrishnan, 2016; Shibaji Bose, 2018).

A granular understanding of material and cultural meanings linked to the resource by communities is therefore imminent. The varied challenges of our cities (many of which are global) and the attendant responses (most of which are local) drive home the fallacies of any monolithic knowledge system like planetary urbanization. Moreover, cities in India vary significantly in terms of geography, politics, social structures, urban fabric, cultural traditions, heritage and economy. This volume establishes the emergent resilience paradigm based on detailed case studies in three medium historic cities in India woven together with generalised observations around this geographical spatiality (of medium historic cities). We agree with Brenner and Schmid (2015) that 'a new vocabulary of urbanization is urgently required that would help us, both analytically and cartographically, to decipher the differentiated and rapidly mutating landscapes of urbanization that are today being produced across the planet' (pp. 176). The argument developed in this volume seeks to serve a starting point for new thinking and responses on urban resilience anchored to the geographical spatiality of rapidly urbanizing medium historic cities in India.

Section 3: Research Methods

The study mainly covers historic medium towns/cities of India that represent the emergent urban reality and serve as laboratories for major transformations, both physical and cultural. Each subsequent chapter undertakes a detailed case study, representing a typology of resilience. A comparative method is used to arrive at this typology. For instance, for water resilience, the primary case is Bhopal with comparisons and secondary references to other cities that represent this typology like Kolkata. Urban form is observed as the tangible product of the relationships between humans and their environment – social, economic and cultural. There is much common ground between the way urban morphologists see, study and conceptualise the form of cities and the way ecologists do the same with ecosystems, at the heart of which is the concept of resilience.

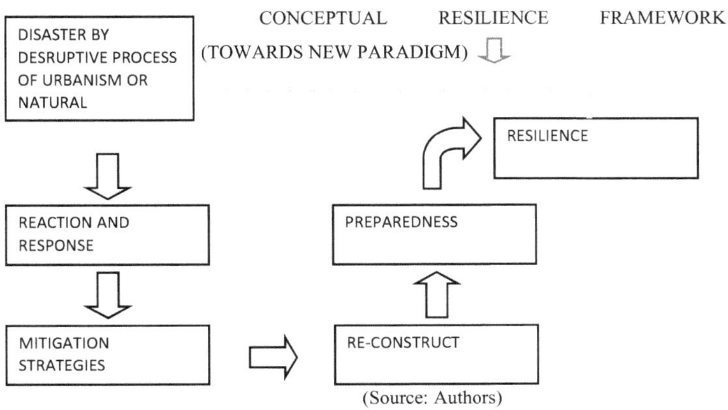

Figure 1.1 Conceptual resilience framework.

This volume is the culmination of detailed ground research in medium historic towns/cities of India. The research questions that form the bedrock of this study are:

1 What are the traditional values and systems that were resilient in each of the chosen cases? (historical perspective)
2 How do they represent a particular typology of resilience–cultural heritage, water, coastal ecology? (this will help us to arrive at the multiple dimensions of resilience)
3 What are the features of this typology that in turn help us to theorize on southern urbanism?
4 In what ways could these traditional values and systems adapt in contemporary times?
5 What are the vulnerabilities posed by current urban development models in these cities, for instance, Smart City plans and projects?
6 What are the resilient strategies, all may not be good (read sustainable), that must be given up, and how could the good (read sustainable) practices be conserved, preserved, designed and redesigned (essentially think of solutions) from design and cultural practice perspectives?

Section 4: Significance of the Study

The study unravels parallel contradictory processes that are currently the hallmark of urbanisation in the Global South, especially rapidly urbanising smaller cities, specifically witnessed in the medium historic towns of India. These are laboratories of smart experiments (for details, see Singh & Parmar, 2019) and are subject to what Tripodi (2020) calls 'technological urbiquity as a neologism to describe a fast-emerging techno-social condition that facilitates remote access to almost everything that once was the reason for the city to exist and endorses the process of planetary urbanization' (p. 2). The use of smart technologies, artificial intelligence, robotics, machine learning, dashboard governance and cities run-on real-time data point towards this trend. Tripodi elaborates further

> 'that it is a condition that has a disruptive capacity in respect to consolidated geographical categories and social roles, but that also obscures the emerging techno-political order and its inherent fragility and unsustainability uses a related metaphor "telescoping the city" to understand the epoch-making shift in the way we perceive and position ourselves in planetary spatiality'.

This process of telescoping the city with the help of new technologies misses out on the story of the life worlds – the crucial interconnections between place, people and culture that can only be understood through the everyday urbanisms of the city at the eye level (for details, see Singh, 2019; Unhale & Singh, 2020). It is here that resilience is also anchored. Understanding the importance of culture in global debates on urbanisation, the Habitat II City Summit in 1996 linked culture with people's well-being, local development, equity and diversity in heritages and values. The Global Taskforce of Local and Regional Governments (facilitated by UCLG) acknowledges the need to explicitly include culture in the success of sustainable development policies, as without culture, there is no future for cities. Cities need vitality, meaning, identity and innovation, and citizens need to widen their freedoms (Source: Global Taskforce of Local and Regional Governments, 2016). Later, the #culture2015goal campaign pointed out that national strategies need to be adapted to their cultural context to be effective and that a broad understanding of skills and capacities should prevail. This includes the ability of key stakeholders to be sensitive to cultural aspects, recognise cultural diversity and heritage and embrace creativity (for details, see www.finditnearme.co/resources/culture-2015-goal.html).

Box 1.1 A historical overview of understanding resilience

Natural processes – water cycle, wind at various levels of the atmosphere, movement of crustal plates – and events associated with them – floods, droughts, cyclonic or wind storms, earthquakes – have happened from time memorial on earth and do not directly cause disasters. It is human beings, with limited knowledge of natural processes and either interfering with or ignoring them in their inexorable quest for development, who are directly responsible for disasters. Paradoxically, as human beings have increased their knowledge of the natural processes, they have interfered more with them, and the consequence is becoming increasingly self-evident in the form of climate change that is increasing the frequency of the aforementioned events leading to more frequent disasters. Cities, a dense conglomeration of human beings in one region, have always come up close to water. As urbanisation has increased over the last two centuries due to economic development becoming more concentrated in cities, the risk of disasters has increased significantly, especially as they interrupt the natural water cycle and use up natural resources at an unsustainable rate. Human beings have understood from past disasters (e.g. destruction of Pompeii from the volcanic eruption of Mount Vesuvius) that disaster risk needed to be reduced through trial and error, and, therefore, certain risk-reduction procedures were included in community social norms, which people could follow without understanding the reasons behind these norms (e.g. house construction in Japan from the 1600s with wood and paper to reduce earthquake risk, and houses on stilts in South and Southeast Asia to reduce flood risks). Post– World War II, technology development accelerated in Europe and USA, and risk of disasters was understood to depend on two factors, occurrence rates of natural events and ability of a system to withstand those events. Thus, the risk of disasters could be reduced in one of two ways: one was to reduce the occurrence rate of events and the other was to increase the ability of the system to withstand those events. In cases such as floods, droughts or landslides, the event occurrence rates could be reduced using engineering solutions. However, in cases such as earthquakes and wind storms, occurrence rates cannot be reduced, but prior information on occurrences of events can reduce risk. Increasing the ability of a system, such as a building or a community, to withstand events can be done by incorporating engineering mitigation measures. However, despite great strides in engineered disaster risk reduction and maybe due to them, disasters continued to occur, primarily as rapid urbanisation occurred ignoring risks and forgetting community social norms (considering them to be silly superstitions). Thus, disaster management techniques were introduced to manage systems post events to ensure the communities were able to return to normal activities quickly. Institutions were created in the USA (Federal Emergency Management Agency in 1979) and other OECD countries in the 1980s to manage disasters, especially due to floods, wind storms and earthquakes. Of course, disaster management morphed to include disaster risk reduction as a part of the continuum since late 1980s. In India, the National Disaster Management Authority (NDMA) was created by the Government of India in 2005. Most of humanity lives close to water, and increasing interference with the water cycle in the 20th century had ramifications that scientists started understanding in the last part of the 20th century. The United Nations (UN) created an independent body called the Intergovernmental Panel on Climate Change (IPCC) in 1988. Thus, climate change and its effects were studied, and the correlation between climate change and increasing disaster occurrence rates was identified. As a result of this, the Millennial Developmental Goals were created by United Nations in 2000 which first identified ensuring environmental sustainability as one of the goals to mitigating the effects of climate change and associated disasters, and these morphed into the Sustainable Development Goals (SDG) in 2015. It is interesting to note that engineering disaster management has always been a top-down approach, where all activities are done by governmental agencies at various levels for communities, with no feedback mechanisms from communities even considered. Even in the OECD countries, it is only in the last few years that it is being understood that interaction between government and communities is important, although concrete examples are few and far between. India is a country that is affected by all natural events, except volcanic eruptions, in recent history. The Indo-Gangetic Plains are affected by floods every year; the east coast of the country (and the west coast less frequently) is affected by multiple cyclonic storms every year; the mountains to the north are relatively young and landslides and rockslides are common occurrences, especially during the monsoons; earthquakes occur at the northern boundaries and in the western part of Gujarat at a fairly frequent rate; and droughts are getting more common as the interference of the water cycle is increasing. Thus, disasters are fairly routine, and historically communities developed social mores and changed behaviour to reduce risks – for example, use of wood and light construction material in the north to reduce earthquake risks, sustainable use of wood to ensure that mountain slopes had trees to reduce the risk of landslides, and bamboo houses on stilts in the eastern part of the country – which also had sustainability at their core. From the 1850s, river works were introduced in India, and an extensive canal system created to reduce the flood–drought cycle in Bengal to reduce the flooding due to cyclonic storms. After independence, engineers trained in Indian engineering colleges were at the forefront of river basin management, and large hydroelectric projects were created in the 1950s and 1960s to manage floods and droughts, and also generate electricity for economic development. However, rapid population growth, disuse of a large number of canal systems, and the relative importance of agriculture waning since the 1980s due to growing disparity in rural earnings vis-à-vis urban earnings caused rapid urbanisation and mushrooming of communities in the floodplains of the great rivers and in the tidal plains in coastal areas. Thus, flood and cyclone disasters increased with time. The large hydroelectric projects caused other environmental issues (siltation and inundation of large tracts of arable land) that affected communities in the neighbourhood and upstream of these dams, with the consequence that flood management using dams is no longer considered desirable in India today. It is interesting to note that in India, except for some notable measures that concentrated on prevention of disasters (e.g. the 1970s Chipko movement in Uttarakhand to preserve trees, the Bishnoi conservation movement that started in the 1700s, water conservation in Rajasthan led by Rajendra Singh in the 1970s), almost all disaster management has been run by the government with no community participation even today, with sometimes grave economic consequences to the communities affected by the disaster management protocols. Thus, while disaster risk reduction is in the realm of government even today, sustainability (which gained widespread popularity in the 1970s) was always run by affected communities, with almost no help from the various levels of government. India has been at the forefront of development of the IPCC protocols, but it is only in the last ten years that the government acknowledged that climate change and sustainability are imminent urban challenges. Resilience as a buzzword was probably first used in India after the disastrous flooding of Mumbai in 2005. However, the dichotomy between disaster risk and resilience remains even today – the former the realm of the government and the latter the responsibility of the community.

Author: Prof. Pradipta Banerjee, Director, Centre for Urban Sciences and Engineering, IIT Bombay, Mumbai, India

Section 5: Overview of Chapters

In Chapter 2, we draw from our previous research on Smart City Mission in India that came out as a volume titled *Smart City in India: Urban Laboratory, Paradigm or Trajectory?* published by Routledge in 2019. We extend that research to add more typologies with statistical information, images and maps to drive the point that medium towns in India are the emergent urban reality in India. In this chapter, we discuss in detail the rapid urban transformation that medium towns are currently subjected to both materially and culturally and how these impact their resilience. Chapter 3 carries forward the discussion and shifts focus on historic urban cores and their cultural resilience with a detailed case study of the city of Nashik in Maharashtra. Nashik has emerged as an important economic hub as the region began to develop with the establishment of expressways, Special Economic Zones (SEZs), entertainment zones and second homes. This chapter establishes the cultural resilience typology with the help of the prototype of Nashik and also with secondary references to other cities like Kolhapur and Ujjain. Many of the historic urban cores in India are found to be crumbling under neglect or development pressures. The vulnerability of the cultural heritage in the historic core is related to lifestyle changes, stagnation, decay and outmigration towards new developments in the peripheries and new towns. In spite of these vulnerabilities, communities exhibit inherent resilience when the question is about an object or an act which they value. How does one evaluate cultural heritage – tangible as well as intangible? 'Cultural resilience' entails retaining the identity of the place and communities in a highly globalised world. These identities have been incrementally built by generations before ours. In the context of climate change, the discussions on the ecological impact inherent in the current models of demolition and new construction development are significant. The author subscribes to the idea that the resilience model of conservation and retrofitting of heritage assets to new-age requirements, partially at least, relieve the pressures on our natural resources.

Chapter 4 establishes water resilience with the help of a detailed case study of water systems in the city of Bhopal in Madhya Pradesh. The case study in the city of Bhopal offers a theoretical and spatial framework that could initiate the understanding for resilient and water-sensitive urban development. Historic towns and cities have exhibited complex and innovative water systems to make them self-sufficient over the ages, contributing to bio-geological as well as sociocultural identity through their inherent presence. The concepts of piped water supply, growing urbanisation and impacts of climate change have brought the most renowned cities on the brink of a water war. Even those cities that have water as a resource, face grim scenarios considering they exhibit dismal planning techniques and complete neglect of existing water systems. Further, the wrath of climate change either through droughts or excessive rain make cities and communities vulnerable whereby there is water all around but not a drop to drink. This chapter compares such historic cities (e.g. Kolkata, Jabalpur and Agra that have a close cultural association with water) that exhibit age-old water systems to learn as well as engage with them to make cities resilient in the course of failing times. This chapter further articulates that through water resilient strategies, there is a larger urban paradigm that is addressed.

Chapter 5 represents ecological resilience with a detailed case study tied to the empirical context of the city of Kochi in Kerala with secondary references to similar cities like Panjim and Visakhapatnam. Discourses on urban resilience in the Global South in the present epoch of Anthropocene cannot be limited to human-centric constructs of sociopolitical and economic concerns. The coastal cities of the Indian peninsula are symptomatic of this. These cities have seen an increase in both the frequency of incidence and the intensity of climatic vagaries like coastal flooding, hurricanes and an increased erosion due to increased wave action attributed to the rising sea levels. Interestingly, the past decades have also seen their population increasing steeply as compared to the earlier century. This chapter critically delayers and analyses coastal Kochi, an amphibious port-based agglomeration spread across the islands of the Vembanad estuary, one of the largest estuarine landscapes in the Indian subcontinent. Addressing the city as a rhizomatous hyperconnected network of biotic and abiotic interactions opens up the possibilities of redefining the 'southern socioecological resilience' in a more granular manner. The findings of the study could help in filling the voids in the current 'normative' understandings of resilience which are heavily biased by the discourses of the Global North. In Chapter 6, we summarise and present our conclusive statements by revisiting the research questions. We seek to answer them with the help of each case study conclusion, present the solutions that each chapter draws, draw the typology and theoretical framework that can guide future work and inform policy.

References

Amin, A. (2013, August). Telescopic urbanism and the poor. City: Analysis of urban trends. *Researchgate*. https://doi.org/17.10.1080/13604813.2013.812350.

Baviskar, A. (2007). Indian indigeneities: Adivasi engagements with Hindu nationalism in India. *Indigenous Experience Today*, 275–303.

Benjamin, S. (2008). Occupancy urbanism: Radicalizing politics and economy beyond policy and programs'. *International Journal of Urban and Regional Research. Researchgate*, 719–729.

Brenner, N., & Schmid, C. (2015). Towards a new epistemology of the urban? *City*, 19(2/3), *Academia*, 151–182.

D'Souza, R., & Nagendra, H. (2011). Changes in public commons as a consequence of urbanization: The Agara lake in Bangalore, India. *Environmental Management*, 47(5), 840–850.

Ernstson, H. V. (2010). Urban transitions: On urban resilience and human-dominated ecosystems. *AMBIO*, 39(8), 531–545.

Folke, C. C. et al. (2002). Resilience and sustainable development: Building adaptive capacity in a world of transformations. *AMBIO: A Journal of the Human Environment*, 31(5), 437–440.

Friedmann, J. (2010). Place and place-making in cities: A global perspective. *Planning Theory & Practice. Researchgate*, 149–165. https://doi.org/10.1080/14649351003759573.

Ghertner, A. (2010). Calculating without numbers: Aesthetic governmentality in Delhi's slums. *Economy and Society*, 39(2), 185–217. https://doi.org/10.1080/03085141003620147.

Gidwani, V., & Baviskar, A. (2011). Urban commons. *Economic & Political Weekly*, 46(50).

Goldman, M. (2011). Speculative urbanism and the making of the next world city. *International Journal of Urban and Regional Research*, 35(3), 555–581.

Holston, J. (2008). Insurgent citizenship: Disjunctions of democracy and modernity in Brazil. *The Journal of Latin American and Caribbean Anthropology*, 14(2), 520–522.

Leichenko, R. (2011). Climate change and urban resilience. *Current Opinion in Environmental Sustainability*, 3(3), 164–168.

Mukherjee, J. (2017). *Sustainable Urbanization in India: Challenges and Opportunities (Exploring Urban Change in South Asia)*. Springer, Singapore.

Nagendra, H. (2016). *Nature in the City: Bengaluru in the Past, Present, and Future*. Oxford: Oxford University Press.

Ostrom, E., & Nagendra, H. (2014). Applying the social-ecological system framework to the diagnosis of urban lake commons in Bangalore, India. *Ecology and Society*, 19(2).

Parthasarathy, D. (2011). Hunters, gatherers and foragers in a metropolis: Commonizing the private and public in Mumbai. *Economic and Political Weekly*, 46(50), 54–63.

Robinson, J. (2006). *Ordinary Cities: Between Modernity and Development*. Routledge, London: New York.

Rómice, O., Porta, S., & Feliciotti, A. (2020). *Masterplanning for Change: Designing the Resilient City* (1st ed.). RIBA Publishing, London, https://doi.org/10.4324/9781003021490.

Roy, A. (2011a). Conclusion: Postcolonial urbanism: Speed, hysteria, mass dreams. In A. O. Ananya Roy (Ed.), *Worlding Cities: Asian Experiments and the Art of being Global* (pp. 307–335). Wiley-Blackwell. Malden, MA.

Roy, A. (2011b). Slumdog cities: Rethinking subaltern urbanism. *International Journal of Urban and Regional Research,* 35(2), 223–238.

Roy, A., & Ong, A. (2011). *Worlding Cities: Asian Experiments and the Art of Being Global.* Wiley-Blackwell. Malden, MA.

Schindler, S. (2014). A New Delhi every day: Multiplicities of governance regimes in a transforming metropolis. *Urban Geography,* 35(3), 402–419.

Schindler, S. (2017). Towards a paradigm of southern urbanism. *City,* 21(2). https://doi.org/10.1080/13604813.2016.1263494, 47–64.

Seema Mundoli, B. Manjunatha & Harini Nagendra (2017). Commons that provide: the importance of Bengaluru's wooded groves for urban resilience. *International Journal of Urban Sustainable Development,* 9:2, 184–206, DOI: 10.1080/19463138.2016.1264404.

Sen, A., & Nagendra, H. (2018). Imperilled waterscapes: The social-ecological transformation of lakes in Bengaluru. *Ecology, Economy and Society – the INSEE Journal,* 3(2), 125–134.

Shail, M. (2009). *The Other Global City*. Routledge, New York.

Shibaji Bose, H. K. (2018). Uncertainties and vulnerabilities among the Koli fishers in Mumbai: A photo voice study in culture, place and ecology, special issue edited by D Parthasarathy and Binti Singh. *Indian Anthropologist,* 48(2).

Simone, A. (2001). On the worlding of African cities. *African Studies Review,* 44(2), *Ways of Seeing: Beyond the New Nativism,* 15–41.

Simone, A. (2014). *Indonesia. Jakarta: Drawing the City Near*. University of Minnesota Press, Minnesota.

Singh, B. (2018). *Divided City, The: Ideological And Policy Contestations in Contemporary Urban India*. World Scientific, Singapore.

Singh, B. (2019). Citizens, Spatial Practices and Resurrection of the Idea of Place in Contemporary Lucknow. In *Interdisciplinary Unsettlings of Place and Space* (pp. 165–181). Springer, Singapore.

Singh, B., & Parmar, M. (2019). *Smart City in India: Urban Laboratory, Paradigm or Trajectory?* Routledge, London.

Srinivas, S. (2004). *Landscapes of Urban Memory: The Sacred and the Civic in India's High-tech City*. Orient Blackswan.

Srivastava, S. (2014). *Entangled Urbanism: Slum, Gated Community, and Shopping Mall in Delhi and Gurgaon*. Oxford University Press.

The Global Taskforce of Local and Regional Governments. Roadmap for Localising the SDGs: Implementation and Monitoring at Subnational Level; UCLG: Barcelona, Spain, 2016.

Tim Edensor, M. J. (2012). *Urban Theory Beyond the West: A World of Cities*. Routledge, London: New York. ISBN 9780415589765.

Tripodi, L. (2020). Telescoping the city: Technological urbiquity, or perceiving ourselves from the above. *Space and Culture,* 23(4).

UN Habitat. (2016). Retrieved from Urban Crises Charter. Global Alliance for Urban Crises WHS Brief: https://unhabitat.org/sites/default/files/download-manager-files/Global%20Alliance%20for%20Urban%20Crises%20WHS%20Brief%20-%20Final.pdf

Unhale, S., & Singh, B. (2020). *How Will India Fix Her Urban Future? Between Architecture and Urbanism,* Mumbai.

Unnikrishnan, H. (2016). Contested urban commons: Mapping the transition of a lake to a sports stadium in Bangalore. *International Journal of the Commons*, 10(01), 265–293.

Unnikrishnan, H., & Nagendra, H. (2014). Privatizing the commons: Impact on ecosystem services in Bangalore's lakes. *Urban Ecosystems,* 18(2).

2

MEDIUM HISTORIC TOWNS

The Emerging Urban Reality in India

Manoj Parmar and Binti Singh

India does not live in villages anymore. India also does not live in big cities anymore. Nowhere is this urban transition more visible than the medium historic towns of India. These include Jaipur, Indore, Bhopal, Madurai, Mysore, Coimbatore, Ujjain, Agra, Varanasi and many more medium-sized towns that are beyond the large metro's peripheries, yet impetus or forces of metros are clearly visible in their transition.

Medium historic towns of India like Madurai, Bhopal and Jaipur have always been in transition and often challenge their erstwhile economic production systems and social life. The contemporary nature of medium historic towns manifests the palimpsest of the past through its urban character and other artefacts. There are several medium historic towns in India that are still living cities, keeping age-old customs and rituals simultaneously with new means and ways of producing architecture and livelihoods. These along with several other urban centres across India have metamorphosised over the years into sacred cities, capital cities of medieval times, colonial cities developed through trade and later as administrative capitals and post-colonial cities. Each layer is directed towards betterment in city administration and governance vis-à-vis the existing old city. The medium historic towns of India witnessed massive transformation in the colonial era wherein

 DOI: 10.4324/9781003098461-2

new infrastructure, amenities and institutions were overlaid on the existing fabric as an exercise towards consolidation of political power. The post-1990s urban restructuring has accelerated with economic liberalisation and concomitant changes in institutions, governance, culture and society established in literature (Nijman, 2006, 2007; Singh, 2014; Chaplin, 2007; Shaw, 1999). Archetypical conflicts and contestations traversing the environment, livelihood, depleting liveability, housing and affordability are also essential accompaniments of the rapidly urbanising medium historic towns.

The Smart City Mission launched in 2015 aims to transform selected medium historic towns across the country into models of technological and infrastructural innovation. Under the wing of the Ministry of Housing and Urban Affairs, GOI, every Smart City is supposed to feature housing for all, comprehensive public transport, green spaces, walkable streets, dependable water, electricity and internet connectivity, and citizen-friendly governance, everything digitally controlled and administered. The normative implications of such mass overlaying of digital infrastructure over existing historic cities have far-reaching consequences. The three layers of urban transformative imperatives that work simultaneously on all medium historic towns are:

1 Indian urban planning modalities on satellite cities, industrial cities, knowledge cities and SEZs
2 Global urbanism imperatives on ecological urbanism, resilient cities and climate change
3 Smart City Mission and consequent urbanism that aims to achieve real-time networked cities and dashboard governance

The emerging challenges of the three layers of urban design/planning framework may not necessarily have anything in common. There is a conspicuous absence of any structured discourse to find common parameters and inter-relationships. The challenges that result from this disjunction in the existing planning paradigm have excluded the primordial and often organic necessities of medium historic towns of India. The discourse on Indian urbanism can be traced from the multiple layers of history. Each layer has left behind a rich repository of architecture, urban phenomena and everyday urban life. The Smart City Mission (SCM) has also ushered in new possibilities for these cities often neglected in the urban narrative of India. However, examination of ground realities presents haphazard patterns, fragmented spaces and random initiatives in the process of urban planning and physical transformation. These random smart experiments are often not connected to the complex urban realities especially as witnessed in medium historic towns.

Section 2 is a detailed discussion on the urban transformation processes currently witnessed in the medium historic towns in India. Sections 3–8 are detailed descriptive accounts of various categories of medium historic towns. It is a daunting task to be able to categorise all such urban centres within the geography of India. Notwithstanding, we undertook this task to place them in conceptual categories that we think are nearest to their fundamental characteristics. Future research may disagree with or modify these conceptual categories. However, our examination of current urban realities suggests that these categories offer a holistic and easy understanding of the medium historic towns of contemporary India. Section 9 revisits those erstwhile medium historic towns that were once significant urban centres but have steadily declined and perhaps remained as relics of the past. Yet, they offer an important understanding of the conditions and phases of urbanisation in India that provided the impetus for their growth and prominence. Section 10 concludes this chapter by summarising the key points, presenting the conclusive remarks and linking with the discussion in the next chapter.

Section 2: Urban Transformation in Medium Historic Towns in Contemporary India

The term second-tier city as per the Urban and Regional Development Plans Formulations and Implementations (URDPFI) guidelines (see Table 2.1) applies to cities with a population of less than one million for its administrative structure and along with those, cities that are regional hubs such as state capitals or industrialised centres. The term also implies the nature and characteristics of such cities with respect to their mobility networks, infrastructure, social and cultural institutions. Incidentally, most of the second-tier cities as per URDPFI guidelines are historic cities of India with rich cultural and architectural heritage and often such cities also carry religious significance. Second-tier cities are classified as medium towns in two categories based on population of 50,000–100,000 as medium town I and 100,000–500,000 as medium town II. Both types are administered by municipal councils. In India, within the category of medium town II, there are 372 cities. Indian urbanisation trends reflect global trends of higher income groups correlated to urbanised areas. However, the Indian story of urbanisation is also a story of environmental degradation, pollution, high densities and compromised quality of life.

The first category of smaller towns is referred as 'transitional towns' in the 74th Constitutional Amendment Act (CAA) 1992, where an administrative structure of Nagar Panchayat (as a municipality) is to be formed for an area in transition from a rural area to an urban area. There are more than 7,000 towns under such categories with similar attributes of historicity and a traditional pattern of evolution to medium town.

Table 2.1 Classification of Towns and Cities

Classification	*Sub-category*	*Population Range*	*Number of Cities as per the Census of India 2011*
Small town*	Small town I	5,000–20,000	
	Small town II	20,000–50,000	7,467
Medium town	Medium town I	50,000–100,000	
	Medium town II	100,000–500,000	372
Large city		500,000–1,000,000	43
Metropolitan city	Metropolitan city I	10 lakhs–50 lakhs	45
	Metropolitan city II	50 lakhs–1 crore	05
Megapolis		More than 1 crore	03

Source: Urban and Regional Development Plans Formulations and Implementations (URDPFI), 2014

The URDPFI guidelines talk about the process of planning and transformational nuances of statutory and non-statutory towns with various types of plans, namely Perspective Plan, Regional Plan, Development Plan, Master Plan, Local Area Plan, Special Purpose Plan and Annual Plan. These types of planning processes are imagined to be readdressing, reimagining and resolving the gaps within the planning paradigms. All planning processes are meant to uplift socio-economic conditions, improve accessibility and networks and development of towns, except the Special Purpose Plan that focuses on urban renewal, sustainability, environmental and heritage protection-related issues. The Special Purpose Plans are initiated based on urgency and circumstantial priority for addressing specific issues in a given situation. In addition, these various types of plans are imagined to have interrelationships, directly or indirectly affecting other types of plans in more ways than one.

However, the urbanisation trends and planning paradigms in the Indian context show transformation patterns that often contradict with the envisioned Perspective Plan or Development Plan. The latter sharply contradict with the Regional Plan and often do not provide any opportunity for implementation of the Local Area Plan. The various plans also often do not work in convergence due to sharp contrasts in urban and rural areas. Notwithstanding the fact that there may be several administrative limitations that impede the working of various plans in convergence, this has serious implications of achieving urban liveability index as stated in the Sustainable Development Goals (SDGs).

The URDPFI guidelines also talk about the emergence of Structure Plan as a response to the various limitations stated previously. The Structure Plan is a multiscalar planning tool that operates, directs and creates a framework

Table 2.2 Planning Systems, Scope and Time Frame

Planning System	*Scope and Purpose of the Plan*	*Time Frame**
Perspective Plan	To develop vision and provide a policy framework for urban and regional development and further detailing.	20–30 years
Regional Plan	To identify the region and regional resources for development within which settlement (urban and rural) plan to be prepared and regulated by DPC.	20 years
Development Plan	To prepare comprehensive Development Plan for urban areas, Peri-urban areas under control of Development authority/ Metropolitan Planning Committee.	20–30 years
Local Area Plan	To detail the sub-city land use plan and integration with urban infrastructure, mobility and services.	5–20 years

Source: Urban and Regional Development Plans Formulations and Implementations (URDPFI), 2014

for the development from region, city scale, areas and precinct. The Structure Plan offers an opportunity to integrate various plans optimising on time and resources. The larger discussion in the book around case studies emphasises the emergence of Structure Plan as a future path for various historic cities in India.

The cultural transition of medium historic towns and cities in India is particularly visible in growing consumption patterns of a variety of goods and commodities earlier unheard of. In an interesting article Mofussil.com in the Economic Times magazine dated 21–27 January 2018, one finds reference to the numerous ecommerce companies that are catering to the burgeoning demands of these faraway places. The article mentions that Amazon executives say they get 70% of their business from small towns; the company has set up dozens of new distribution centres. Its arch rival, Flipkart and other ecommerce companies have worked out the intricacies of sending millions of shipments to customers in smaller towns and cities, chasing down the next 100 million internet users in small and medium towns. It is evident that these have emerged as big markets for e-commerce companies in recent years. This also signals the widespread reach of Digital India and the power of the internet, satellite television and new modes of telecommunications. These technologies have spread new awareness and exposure to trends, and created demands and a new class of consumers. These cities also constitute the major bulk audience of the Hindi television and entertainment industries. The huge success of films like *Toilet Ek Prem Katha, Mukkebaaz, Tanu Weds Manu* – all rooted, scripted and filmed in the not-so-palatable small and medium towns of India is one indicator of this trend. Apart from their social relevance, these films have significantly helped bring the small and medium towns back to mainstream popular culture. The growing Television Rating Points (TRPs) of television shows narrating stories from these geographies is also an indicator of how consumption patterns and lifestyles of the hinterland are important determinants of content for the entertainment industries tucked away in faraway big cities like Mumbai. A large part of India resides outside the big cities of Mumbai, Delhi, Kolkata and Chennai. The social groups and communities living in medium historic towns have significantly different entertainment sensibilities and preferences. They have the appetite for consumption and for brands but they want their entertainment content to be delivered in a format and language that appeals to them. The combination of money spreading outside of metros, the rise of a new generation of consumers with different educational backgrounds, emergence of new cultural icons like Mahendra Singh Dhoni, Kapil Sharma and YoYo Honey Singh more rooted in the vernacular has led to massive changes in consumption of media and entertainment.

So, the story of urban transformation of medium historic towns in India can be articulated from multiple points – changing ways of life reflected in the interrelationships of social, cultural practices with natural ecologies, livelihoods and changing association between the natural and artificial artefacts in cities.

Section 3: Planned Historic Towns and Cities

Historic cities in India, their genesis, can be deciphered using simple criteria-trade cities, spiritual cities, planned cities and administrative cities. Each category shows the trends and growth of urban centres with peculiar characteristics.

Jaipur and Fateh-pur-Sikri are examples where cities are known to have aesthetic and architectural compositional qualities along with civic architecture for governance. For instance, the planning principles based on *Vastu Purusha Mandala*, an ancient knowledge system and principles of planning of early Indian cities (a geometric pattern representing the cosmos, often associated with Jaipur) demonstrates the formation of urban form that stands in sharp contrast to the dense, irregular urban grid of medieval cities.

Figure 2.1 Image of Bhubaneshwar city.

Source: Authors

Figure 2.2 Streets in the city of Madurai.

Source: Authors

The planning and transformation of Jaipur brings important urban learnings from various perspectives. Jaipur exemplifies the intrinsic relationship among various components that make and shape the culture and social life of the city. Perhaps, it is one of the prime examples.

Madurai and Seringapatam are other examples of old cities in South India that emanate from sacred centres to advanced systems of planning of quadrants with respect to occupation and social backgrounds. Madurai grew out of layered cosmic diagrams, depicting various realms with respect to sanctity and spirituality of the sacred. The resultant urban form that grew out of it into four distinct quadrants, reinforce the layered nature of realms – sacred and public. Contemporary Madurai, in the throes of transformation faces several issues related to economic sustenance, spatial divisions, basic infrastructure and amenities posing serious challenges for its future and moving towards the sustainable development goals.

Section 4: Urban Centres of Religious Importance

The next categories of urban centres are those of religious importance. These grew out of their sacred geographies, or mythological presence

Figure 2.3 Ghats of Haridwar.

Source: Authors

Figure 2.4 Ghats of Nashik.
Source: Authors

in scriptures, historical texts and source, legends, folklore and local wisdom.

The cities especially along river fronts, natural or artificial lakes are medium towns of category I. Along with that, cities of religious significance as places for preaching wisdom and enlightenment are designated as small town II as per URDPFI definition. The cities of Nashik, Ujjain, Kanchipuram, Varanasi, Amritsar, Ajmer, Nanded, Pandharpur, Gaya, Sarnath, Mathura, Trimbakeshwar, Jabalpur and Jodhpur are such old historic cities whose genesis of evolution run deep in their mythology and sacred geographies.

All such cities are living cities and continue to practise age-old rituals and traditions. They also face a common challenge of rapid depletion of value embedded in the sacred nature of living practised for centuries. Such cities continue to grow alongside new means of production and manifestation, while keeping the historic values and rituals more or less intact. These cities also have visible layers added to their architecture over the years and transformed urban form.

The cases of Jabalpur and Jodhpur are important because of their relationship with water. Water bodies are physically and spiritually integral parts of several historic cities of India. The presence of water bodies within the cities not only add symbolic value but also address the water needs of the city. Water as an urban system in the context of supply and waste water goes beyond engineering to evolve as a sociological–ecological system that forms an intricate relationship with the community and the city. However, these cities are constantly transforming themselves through newer means of planning, newer land use and newer relationships with the ecological system that often conflict with each other. Resultantly, the physical and spiritual essence of water continually degrades over a time period. The once privileged position of water bodies in such historic cities are subjected to land formation by landfill or are ignored as residual components, systematic encroachment of edges and deterioration of systemic relationship of water, community and the city.

Jodhpur, Bhopal and Jabalpur are a few examples out of many medium historic towns and cities across India, where historic relationships (between ecology, people and culture) exhibit capabilities to adapt to newer challenges through urban water resilience strategies. The latter got disrupted, over time, with serious implications on climate resilience. The effectiveness of a water-based resilient infrastructure or its responsive urban fabric and architecture depends on its ability to anticipate, absorb, adapt to and/or rapidly recover from a potentially disruptive state and is even able to return to its original state. A thorough understanding of the historic water system of such cities is therefore imminent to build urban resilience, a matter of concern in contemporary times and requires concerted efforts from academia and policymakers.

The city of Nashik is another example where one can witness how cities have evolved and the ways in which its rituals, social practices and cultural values are embedded in its urban fabric. This aspect of the city is in sharp contrast with contemporary planning practices that are pushing the city to grow outward and turn away from its river and treat the historical ghats as

Figure 2.5 Image of the city of Jabalpur.

Source: Authors

Figure 2.6 Image of Shimla city.
Source: Authors

backyards of depletion. History of urbanisation of such cities demonstrates that culture is also a key to several urban transformative processes, enabling the formation of key civic architectural precincts, monuments, heritage, traditions and a vibrant public realm. Public realm of these cities is the mainstay through which the social and cultural constituents manifest and form the historic urban fabric. It is this unique historic urban fabric formed over centuries that renders place-based cultural identity to the city and requires immediate attention before it succumbs to global homogenisation.

Section 5: Urban Centres of Cultural Importance

Urban heritage and culture have always been neglected areas in cities of India. Successive empires and dynasties have added to the complexities of the urban landscape that still reflect in the built environment and strong presence of local traditions. Many are worth revisiting, learning from and above all conserved. Cultural practices, customs, rituals and traditions are recognised as essential components of each city's place-based cultural identity. The city of Lucknow, Kochi, Hyderabad and Junagadh are some of the important cities that grew out of commerce, trade or administrative intent and have become important centres with intrinsic cultural values manifested in social practices, architecture, street and public realm and indigenous produce over time. The place-ness condition is incontestably a repository of distinctive culture. The global report on culture for sustainable urban development by UNESCO (Duxbury, 2016) brings centres of culture in the sustainable model for urban development deliberating on key questions like: What is the role of culture in urban development? How has culture influenced urban development across the world? How can culture make a difference for our urban future?

Cities worldwide respond to global homogenisation differently and often they mirror themselves as reflected images of large cities. Jenifer Robinson in her scholarly work Ordinary Cities (2013) hints at growing scholarship and importance of southern urbanism in order to develop a new urban theory that is non-hierarchical and beyond developmentalism. The cultural discourse is re-emerging as a symbiotic relationship between place, culture and economy (Singh, 2018). Cultural expressions give people the opportunity to identify them collectively, to read the traces of history, to understand the importance of traditions for their daily life, or to manifest realities. The question is how do we translate these fundamental social and human needs already embedded

Figure 2.7 Image of Kochi city.

Source: Authors

in the urban fabric into sustainable urban planning? It is necessary to bring forth the cultural discourses in the form of cultural theory and cultural studies so that the cultural argument can be centred in a comprehensive resilience discourse. The latter needs to encompass cultural territorial aspects of public realm, activity patterns along with history and heritage to enable the making of a cultural sustainability system integrated with existing planning mechanisms.

Section 6: New Greenfield Urban Centres

Recent times have seen a shift in urban planning concerns in India with the SCM driving urban centres to move from functional, hierarchical forms to networked forms. The cities that are categorised under the SCM cover a wide spectrum of greenfield sites like Dholera, GIFT and Amravati. The planning paradigm that is imagined under Special Economic Zones (SEZ) fosters greenfield development along various economic corridors like the Delhi Mumbai Industrial Corridor (DMIC), connecting various production-based cities (like Dahej and Sonipat), with a feeder corridor to export goods. The third category of greenfield urban transformation has led to the rise of utopian suburbs for urban *bourgeoisie* in exclusive private living environments like Lavasa and Amby Valley. This category stands on consumption and lifestyle patterns suited to the global audience. Greenfield development poses serious environmental and social challenges – eking out of the natural ecologies and resources of the hinterland, built around high consumption of resources for private needs. These have serious implications on questions of resilience and inclusion. These developments pave the way for planning

Figure 2.8 Image of Dholera.

Source: Authors

(based on neo-rational epistemology) that pose a major threat to inbuilt resilient strategies of many of our organic geographies.

Section 7: Leisure Cities

The cities that fall under these categories are the most neglected in mainstream urbanism in India. The core strength of leisure cities revolves around ecology and tourism. Kodaikanal, Ooty in the south, Darjeeling and Gangtok in the east, Shimla, Nainital, Mussoorie and Dalhousie in the north and Mahabaleshwar and Matheran in the west are falling prey to rampant and unchecked development evading many ecological concerns.

The city of Shimla has a unique topography and is one of the few cities in the world that has two urban forests within the city limit. The last several decades have witnessed glaring transformations in social life that have engulfed the entire topocentric urbanism of Shimla.

Shimla has seen political transformation from summer capital in British India to the capital of Punjab and now the capital of Himachal Pradesh.

The topocentric imagination of present living defies the conventional norms to comprehend the situation. The entire terrain is separated by means of roads

Figure 2.9 Image of Mussoorie.

Source: Authors

that occupy the plains and splits the terrain innocently chiseled out of a monolith. The virtue of 'what is built' and 'what is left' is carved reality of a displaced collective. The case of Ooty, Mahabaleshwar or Nainital is no different from Shimla. A *Laissez-faire* growth catering to increasing demands of tourism is throwing existing planning mechanisms out of gear. One really needs to assess the newer planning goals like SCM and read them along with the existing ecological systems entrenched through the years.

Section 8: New Centres Catering to Educational and Cultural industries

Recent times have also witnessed the rise of newer urban centres across India whose economic base has shifted from industrial/manufacturing production to services. The newer mode of production has also yielded new means of adaptability to existing infrastructure and amenities. There are several medium towns and cities that are transforming with rapid pace like Kota, Hyderabad, Bangalore, Ramoji Film City, Malappuram, Kozhikode, Thrissur and Kollam. These are leaving the fast-changing metropolitan cities behind, as they embark on alternate mode of economic restructuring witnessed over the last decade. For instance, Kota is a city situated on the bank of Chambal River and is the third most populous city in the state of Rajasthan. The development of Kota city is very unique. In recent times, Kota has emerged as a premier educational hub (dotted with numerous coaching centres) that train students for various entrance examinations for professional courses especially into the coveted Indian Institutes of Technology (IITs). Every year over 150,000 students migrate to Kota from various parts of India and enrol in various coaching classes for a one-year period of dedicated study. As the educational-related activity intensifies over the year, the city also witnesses the highest growth of population in the state of Rajasthan. Such spatial–temporal growth and energised economic activities have significantly impacted the city and its peripheral areas. Existing building types transform into mixed-use activities with newer temporal stay accommodations mushroomed across the city and several eateries and cafes catering to young students. The overall land use has significantly altered and put undue stress on the existing infrastructure. The analysis further reveals that significant amount of land transformation on very fertile and productive land has taken place over the past decade.

This information is alarming as well as useful to local government and urban planners aiming for sustainable development of the city. Kota city, which is now an established education hub of India with unprecedented population growth coupled with unplanned developmental activities, has led to rapid urbanisation (Dadhich et al., 2017). These developments pose serious challenges on the environment quality and resource base of the region and require immediate attention. To protect and conserve the arable land and ease the ecological stresses is imminent.

Such transformations can also be seen in Bangalore and Hyderabad labelled as information technology (IT) cities of India that are in constant conversation with research on IT and artificial intelligence (AI). Both the cities enjoy the status of the fastest growing cities in India (third and fourth rank). The city of Bangalore has rapidly changed in the last two decades, with similar characteristics–newer formed economic activities, restructuring of urban areas and its peripheries contributing to stress on existing land use, infrastructure and amenities along with ecology.

Section 9: Decline of Erstwhile Significant Towns and Cities

The emergence of Industrial Towns after independence in the 1950s was purely based on production where criteria for formulating and implementing were based on economic indicators. There are also examples of cities that raised to prominence as manufacturing hubs like Shyamnagar (cotton mills), Titagarh (paper mills), jute mill towns, steel towns, Satna (prism cement), Ichalkaranji, Bhiwandi (textiles) and Chandrapur (mines). In this category, the former towns were planned while later ones organically emerged with the passage of time. However, in both cases, the common denominator was economic growth. Cities either programmed themselves to evolve new infrastructure or managed through *laissez-faire* growth. The later ones capitalised on the de-industrialisation policy of existing metros and emerged as alternative manufacturing hubs. Recent urban transformation has created issues on social fronts with these cities witnessing rapid decline in industrial enterprise and subsequent rise of informal growth as backyard warehouses. These were already uniquely positioned with very high densities, labour migration and rising levels of antisocial activities. It is intriguing that there is a clear absence of policy that takes account of social indicators of such erstwhile industrial towns and cities that are increasingly turning defunct in the post-industrial age.

Section 10: Conclusion

The history of urbanisation and contemporary stories are uniquely positioned in Indian context – vast, diverse and heterogeneous. The rationale of this categorisation drives home the imminent need of context-specific urban policy to make such towns develop their own resilient strategies integrating ecological, social and economic indicators.

India as a geography is huge, multifaceted with complex social and cultural diversity, varied economic bases and unique place-based identities. In such situations, the depletion of city fabric or ecological conditions happen with the same intensity as they urbanise. As cities embark on implementing newer planning models they are often treated as objects that require physical face-lifting and makeovers. Such is the case with historic cities like Nashik, Jodhpur and Jabalpur where urban policy and planning are oriented to alter the coordinates of use and do not respond to the operative parameter of place. Similar patterns are underway in religious cities, planned cities or organic cities to accommodate newer economic activities resulting in dilution of land policy and indifference to environmental and ecological conditions, all of which have intensified the depleting process in social, cultural and physical form of these cities.

References

(2014). Retrieved from Urban and Regional Development Plans Formulations and Implementations (URDPFI).

(2018). *Mofussil.com in the Economic Times magazine dated 21–27 January*. (n.d.). In *Smart City in India: Urban Laboratory, Paradigm or Trajectory? Routledge,* London.

Chaplin, S. (2007). Partnerships of hope: New ways of providing sanitation services in urban India. In A. Shaw (Ed.), *Indian Cities in Transition* (pp. 83–103). New Delhi: Orient Longman.

Dadhich, A., Goyal, R., & Dadhich, P. N. (2017). Impact of urbanization on arable land in Kota – a geospatial analysis. *Engineering Sciences International Research Journal,* 5, 1–4.

Duxbury. (2016). *Hosagrahar, N. & Pascual, Why must Culture be at the Heart of Sustainable Urban Development?* Global Report on culture for sustainable urban development by UNESCO.

Nijman, J. (2006). Mumbai's mysterious middle class. *International Journal of Urban and Regional Research*, 30(4), 758–775.

Nijman, J. (2007). Mumbai since liberalisation: The space-economy of India's gateway city. In A. Shaw (Ed.), *Indian Cities in Transition* (pp. 238–259). New Delhi: Orient Longman.

Robinson, J. (2013). Ordinary cities: between modernity and development. Routledge. London: New York.

Shaw, A. (1999). Emerging patterns of urban growth in India. *Economic and Political Weekly,* 34(16/17), 969–978. Retrieved June 10, 2019, from www.jstor.org/stable/4407880

Singh, B. (2014). Urban governance in contemporary India. *Contemporary India*, 4, 89–111.

Singh, B., & Parmar, M. (2019). *Smart City in India: Urban Laboratory, Paradigm or Trajectory?* Routledge, London.

Singh Binti, S. M. (2018). *Divided City, The: Ideological and Policy Contestations in Contemporary Urban India*. World Scientific, Singapore.

3

CULTURAL RESILIENCE OF HISTORIC URBAN CORES

Vikram Pawar

Introduction

The conservation of cultural heritage affected by natural calamities as well as armed conflicts has exposed divergent views on the question of heritage conservation as well as community resilience (Holtorf, 2018; Jigyasu, 2013). Given that conservation and restoration of heritage impacted by disasters in the Global South put undue burden on the already affected communities, Cornelius Holtorf, the UNESCO chair on Heritage Future, proposes the definition of cultural resilience as 'the capability of a cultural system (consisting of cultural processes in relevant communities) to absorb adversity, deal with change and continue to develop' (Holtorf, 2018, p. 237). He further offers that 'disturbances that can be absorbed are not an enemy to be avoided but a partner in the dance of cultural sustainability'. This definition assumes significance as it proposes an acceptance of the disturbance in order to move on, which is the inherent objective of resilience. It also builds upon the idea of 'culture as the fourth pillar of sustainability' first crystallised around the beginning of the millennia (Hawkes, 2001) and finally approved as a policy statement by the Third World Congress of United Cities and Local Governments at their World Summit in Mexico in 2010. The original three pillars of sustainable development when its definition was developed in the 1980s were economic growth, social inclusion and environment balance.

However, in the context of southern cities, where the aspirational trajectories of cities often find themselves at loggerheads with the built heritage especially if it is a living heritage, this definition calls for a serious deliberation. The rapid urbanisation in these cities, speculative nature of real estate development and the politics of resource mismanagement tend to often neglect and eventually erase the tangible heritage, promote gentrification and compromise most of the indicators of cultural values, namely 'articulations of communities' identity, aspirations and/or history; stimulation of community dialogue around quality of life, sustainability and respect

DOI: 10.4324/9781003098461-3

for diversity issues; raising the profile of universal human rights' (Hawkes, 2001, p. 33).

To reiterate the submission in Chapter 1, there is indeed a need to abstain from offering a uniform and monolithic understanding of urbanity, modernity, development and now resilience and engage with the specificities of the cities of the Global South.

Specific to pilgrim towns in India, scholars have noted that planning mechanisms in contemporary practice including the conventional master plan approach have been ineffective in adequately responding to issues related to unauthorised and unplanned growth, strained infrastructure and environmental pollution that characterise urbanisation in religious tourism destinations (Shinde, 2011, 2016). Concerns are also expressed on how 'cultural heritage has been, and will be exposed, to risk over time for the combined action of climate change and the human factor' (Camuffo, 2019).

Further, in the backdrop of India's 100 Smart City Mission 2015, a research paper on cultural resilience and the Smart and Sustainable City (SSC) affirms that SSC concept needs enlargement to include acknowledgement of cultural diversity of communities and engagement of 'chrono-diversity' of surviving fabric. The study also identifies an influence of resilience and sustainability thinking in the discourse on integrated urban (re)development and adaptive reuse of built heritage, effecting a paradigm shift (Clarke et al., 2019). Section 2 discusses vulnerabilities of historic cores and highlights the inbuilt resilience of traditional values and systems. Section 3 is a discussion on what constitutes culture and the traditional values. What are the dynamics which contribute towards the ability to adapt? Why do traditional values and systems require to be resilient in contemporary times? Section 4 discusses the resilient strategies that can be adapted to make the historic native towns more resilient. For this, cultural values need to be identified and acknowledged; threats to these at the local level as well as due to global imperatives need to be understood; and strategies developed can help encourage a holistic development of a diverse, healthy urban living which is rooted in its natural and cultural context. Section 5 is an empirical study of Nashik as a case to understand cultural resilience. Section 6 concludes the chapter by suggesting a cultural resilience typology with secondary references and analogies to other cities like Kolhapur and Ujjain in India.

Section 2: Vulnerabilities of Historic Cores

A World Bank blog recording Giovanni Boccardi, Chief of the Emergency Preparedness and Response Unit for the Culture Sector of UNESCO, on the sidelines of 2018 Understanding Risk Forum, held in Mexico City, calls for much more work to be done to put cultural heritage 'front and center' in the disaster risk management agenda. He offers, 'when there is a trauma, people really need to hold on to their cultural landmarks, their symbols'.

This sentiment need not be restricted to traumas alone. People need to hold on to their cultural symbols also in the case of chronic stresses. In fact, chronic stresses often are not obvious and require sensitive investigations, analysis and deliberations in order to be even noticed. Chronic stresses related to marginalisation, loss of livelihood opportunities, built habitat neglect, decay and dilapidation, strained mobility infrastructure; inadequate infrastructure and amenities; fire preparedness and faulty development trajectories merit inclusion in the resilience discourse. These stresses have resulted in the deterioration of quality of life as well as congestion and abuse of the urban space. Congestion in the historic native cores, poor upkeep of the built form and real estate development pressures exacerbate the vulnerabilities thus contributing to a gnawing and gradual erosion of cultural resilience.

The organic nature of urbanisation in native towns around the historic cores as well as the erstwhile *gaothans* (villages) within the modern urbanisation still retain traditions and values which have been built over time. It would be worthwhile to draw from Daskon's (2010) submission that culture and traditional values strengthen livelihood resilience. She further argues that while the impulse for change may come from external influences, adaptation comes from within, through dynamics which are specific to values of the people.

Section 3: Global Imperatives That Pervade Traditional Values and Systems

In 2002, ICOMOS defined Cultural Heritage (CH) as as an expression of the ways of living developed by a community and passed on from generation to generation, including customs, practices, places, objects, artistic expressions and values. Human endeavour to develop or grow as society has witnessed many manifestations in the past. Some of these have harmonised culture with nature while many others have conflicted with it, giving birth to the phrases 'carving out the new frontier', 'taming nature' signifying the times where culture was threatened by nature. Also significant is the distinction that resilience of groups of human beings is very adaptable as compared to culture which is far more vulnerable to extinction especially where it is in conflict with nature (Maser, 1992).

Last six decades have shown a gradual shift in the global perception of the relationship of culture and nature. With the concerns of climate change, global warming, sea water rise and related catastrophic events which have been witnessed more frequently in the last two decades, there has been a caution against a blind leap into 'the bold new future' and a desire to draw from the pages of traditional wisdom and values. Destruction of the objects, artefacts, knowledge and customs which embody these wisdom and values endanger our ability to be resilient. Simultaneously, new construction related to infrastructure as well as built forms continue unabated. The current methods and practices are consumptive of natural resources like hills, forest, topsoil and water, without being effective in provisioning quality of life to the majority of population in urban south. The loss is much more tangible in the historic cores where the associational values with the cultural objects are relatively stronger. Resilience embedded in the cultural heritage of cities can and should be harnessed for bettering the quality of our cities, thus also contributing to the sustainable development objectives and remaining relevant for the contemporary and future narratives of our cities (Beel et al., 2017; Binder, 2020).

Section 4: Global Imperatives/Contemporary Development Paradigm and Resultant Vulnerabilities

This section focuses on the global imperatives/contemporary development paradigm and resultant vulnerabilities that are caused to the traditional values and system.

Historic fabric is vulnerable to contemporary urban pressures, on a day-to-day basis as well as when there is an exponential increase in the scale of events (festivals) or even when there are sudden shocks. Lack of scientific, empathetic and timely approach towards city growth; lacunae in understanding the complexities of such typologies; and misdirected and ill-informed approaches are making the historic cores more prone to deterioration and cultural erosion. Changes in the construction industry environment and lack of patronage affect opportunities of livelihood, the quality of life as well as the aspirations of future generations of the craftsmen. The financial muscle and markets supporting industrialised products and systems further affect the environment for people engaged in the building craft. There is also a lack of adequate research and training at the grassroots level for these communities in the field of conservation.

Piecemeal and patchwork repair strategies not amounting to a holistic conservation approach contribute towards the rapid deterioration of the built form and hence to the urban fabric. Even the once restored heritage, in the absence of a clear upkeep mandate and compromises made on the crucial periodic maintenance, becomes unstable and deteriorates rapidly after the 'conservation', making the effort thankless and 'futile'. Disconnect between citizens, public heritage and projects is an area of concern. This results in further neglect of heritage in present-day cities of the Global South.

Resilient Strategies and How Can It Become Southern Paradigm

In order to make the historic native towns more resilient, cultural values need to be identified and acknowledged; threats to these at the local level as well as due to global imperatives need to be understood; and strategies developed to encourage a holistic development of a diverse, healthy urban living which is rooted in its natural and cultural context. Thus,

1 Developing a culturally and ecologically sensitive master plan, strategy and action plan for conservation and management of the river, tributaries, ghats (including the sacred pools) and heritage assets
2 An urban mobility plan encouraging pedestrian-friendly zones and streets which also avoids dismantling of historic facades on account of road widening
3 Incentivising restoration as well as adaptive reuse of heritage-built structures and enhancing the heritage values associated with historic precincts

Figure 3.1 Chronic stresses leading to increased vulnerability vis-à-vis a proposed model of how culture and values can strengthen livelihood resilience leading to better adaptation and hence cultural resilience.

Source: Author

Figure 3.1 (Continued)

4 Guidelines for development within historic core which respects the roof lines/scapes; encourages building design and craft traditions in material palette not restricted to concrete, glass and steel
5 Facilitating diversity in citizen participation in ideations, decision-making and management of historic cores

Section 5: Case of Nashik

Urbanisation when contextualised geographically, we get urban spaces along rivers, in the valleys of hilly regions, fortified hill tops/plateaus, around lakes, along the coasts, on plains. Urban spaces can also be studied as places of accumulation of surplus agricultural or natural resources where extraction processing and trade requires organised human agglomerations. These primary drivers of urbanisation are then layered with complexities of institutional structures, political discourses, social amenities and opportunities of cultural exchange. Based on the functions they served, precolonial Indian towns can be seen as administrative, strategic, trade towns, religious centres and places of learning. Out of these, the religious centres which are places of learning as well as strategically located places of trade deserve to be acknowledged as a distinct typology. Primarily this is due to the transitional nature of such cities, especially during festivals and fairs. These characteristics include marked seasonality, corresponding impact on urban infrastructure, presence of class of religious specialists, sacred focus, tourist attractions and as repositories of culture (Bhardwaj, 1994).

Out of the numerous town and city typologies of religious significance in India, this chapter focuses on one that is along the bank of a river. River Godavari (Ganga of South) is the longest river of peninsular India and home to more than 60 million people with at least 13 urban centres across four states including Nashik, the case that is being considered here. For Hindus, Nashik is a major pilgrimage destination and significant for rituals associated with the salvation of a deceased person's soul. The city is well known for being one of the locations of the *Kumbh Mela.*[1] Nashik has evolved as an administrative, institutional as well as industrial hub, and more recently has been part of the 100 Smart City Mission launched in 2015.

It is one of the ideal cases to study the dynamics of contemporary urban development vis-à-vis the culturally rooted relationship of human settlements and river ecologies. The scale of temporality and seasonality of this pilgrim age destination, especially during the Kumbh, its impact on urban infrastructure, presence of religious specialists and sacred focus – all qualify Nashik as a distinct typology of religious urban centre.

Nashik is the fourth largest urban centres in Maharashtra after Mumbai, Pune and Nagpur and also features in the central government list of 100 smart cities. Home to Hindustan Aeronautical Factory as well as many other heavy and medium engineering, automobile industries, the city is an industrial hub. It also anchors the triangular economic growth corridors along with Mumbai and Pune. Blessed with a mountainous terrain and pleasant weather, it is the wine capital of India and a fruit bowl of the state.

Nashik is one of India's sacred ancient cities finding mention in literature as early as circa 250 BC while archaeological evidences suggest settlements during the early Bronze Age (circa 1500–500 BC) (Nashik District Gazetteer). It was along the trade routes connecting Pataliputra (present-day Patna, the capital city of the state of Bihar), the erstwhile capital of Magadha Empire to the ports on the west coast. It prospered as an urban hub during the *Satavahanas* as it was along the important trade route from *Pratisthana* (modern-day Paithan) to *Broach* (Gujarat) and *Sopara* (Maharashtra).

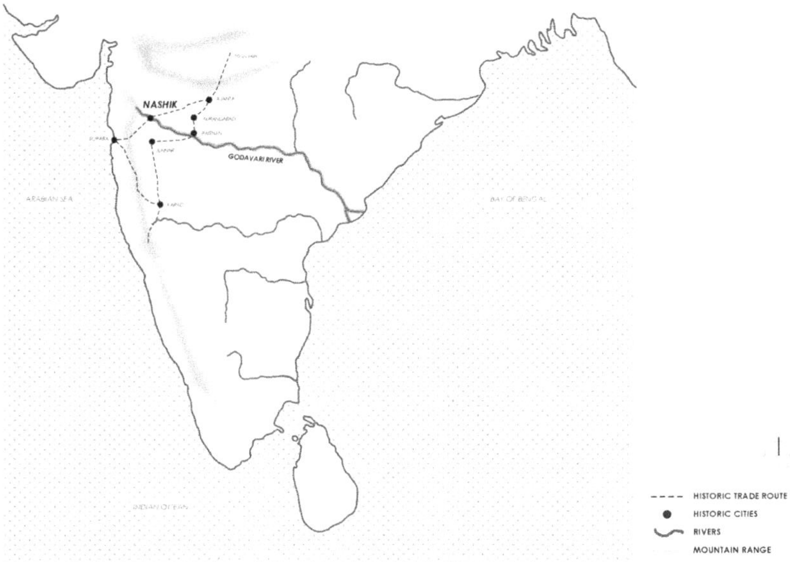

Figure 3.2 Location of Nashik on Godavari, the longest river in peninsular India along the ancient route. Also shown are Nashik's precolonial trade connects with other prominent urban centres of the time.

Source: Nashik Studio KRVIA M. Arch Sem 3, 2019–2021 batch

During the Islamic period, Nashik was named Gulshanabad acknowledging its scenic and pleasant setting. The old name was restored when Peshwas took over in 1751.

Nashik is significant mythologically as well by not only being the abode of Lord Rama's stay during exile but also marking the turning point in the epic where Rama's brother Laxmana chops off Ravana's sister Shurpanakha's nose as a form of punishment for enticing the already married brothers – it is believed that 'Nashik' owes its name to this event. Correspondingly, the sacred river Godavari, considered to be the Ganges of South India, takes a sharp dip southward in its otherwise easterly flow marking an abrupt turning point in the river's flow.

Evolution of the City Around the Historic Core

The historic core of Nashik evolved along the stretch where the river has a southward flow before it resumes its eastern direction. Sixteen recorded

Figure 3.3 Regional context of Nashik showing the drainage pattern, dams and historic core set within urbanised areas.

Source: Nashik Studio KRVIA M. Arch Sem 3, 2019–2021 batch

sacred pools (*kunds*) form the sacred loci along this stretch of the river, the natural bed being disintegrated basalt with intermittent pockets of sand. The upper segment along both sides is believed to have evolved earlier to the Islamic period and called Nava (New)-pura (suburb) suggesting its revival in the Maratha period. The southern/western bank evolved during the Islamic period around the *juni* (old) *gadhi* (fortification). The colonial period saw the expansion outside this historic core and introduction of railway, cantonment town of Deolali and certain planned layouts and irrigation. Diversion of domestic sewage through the east and west canals to treatment plants and water infrastructure project of Gangapur Dam was taken up after independence. Between 1998 and 2004, emergency measures to tackle the Mahakumbh were undertaken resulting in concretisation of the tributaries and the ghats. Since then, there has been an increased public consciousness regarding the negative effects of concretisation of river edges and beds. In 2019, concretisation orders were served by the court and the Nashik Municipal Corporation (NMC) has started its implementation and has restored some of the *kunds*.

The tangible cultural built heritage reflects a transition from the wada typology within fortifications to bazaar house and smaller houses made of masonry and timber to more recent RCC-framed typologies. Some of the oldest surviving architectural objects include restored or newly built temples crafted in stone during the Maratha period.

The built heritage, however, is in a state of neglect. Significant heritage structures are being abandoned, redeveloped for want of better amenities or family litigations or difficulty in upkeep. The *wadas* and the bazaar house typologies (with shops below and residences above) with their robust masonry and intricate/richly ornate woodwork are being permanently lost and have been replaced with generic and poorly articulated RCC-framed typologies thus gradually eroding the identity of the place. The quality of the streetscapes too has been compromised due to partially implemented road-widening programmes to cater for motorised private transport. This has added to congestion, air and noise pollution and led to closure of windows facing the street, redundant balconies, pedestrian-unfriendly streets and poorer quality of urban space.

Ecological Conditions of the City

Nashik is also believed to have derived its name from Nav shikharas or nine peaks (actually spurs extended from the plateau and the number too requires to be validated on site) which channel the courses of two major flows and three minor water streams. At Nashik, the river turns south for

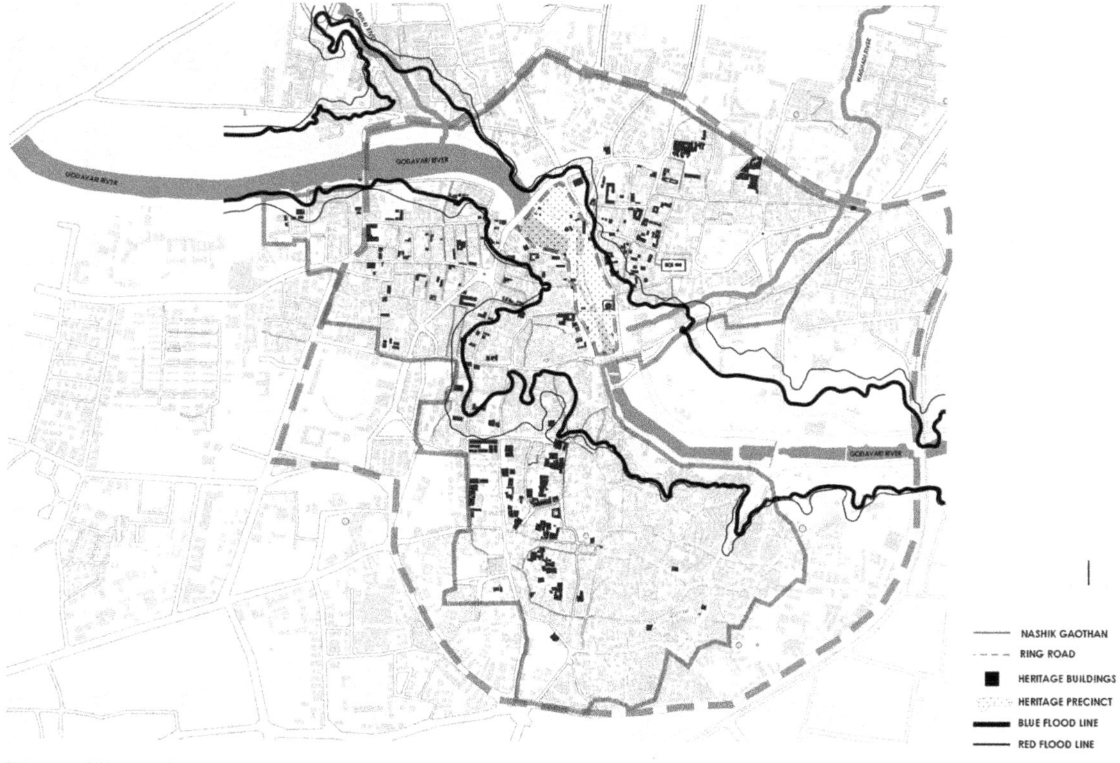

Figure 3.4 The significant heritage within the historic core. The Godavari flows through the centre. The flood lines affect a substantial portion of the historic core since the construction of the dams.

Source: Nashik Studio KRVIA M. Arch Sem 3, 2019–2021 batch

a short stretch seemingly deflected by the mound around Panchvati cave as well as the high velocities of Waghadi which is almost perpendicular to the flow of Godavari. Given the narrowness of the river and lack of depth, these flows also render the opposite bank more vulnerable to flooding. The construction of Gangapur Dam in 1954 also seems to have further aggravated the flood conditions. During the severe rain seasons, when the sluice gates are opened, many of the historic precincts along the bank like Delhi Gate area, now experience floods and the high flood lines demarcated on the Development Plan acknowledge the same.

Storm water and sewage led towards Godavari River and its tributaries. The underground sewage network was first implemented between 1955 and 1968. Between 1998 and 2002, as an emergency measure for the Kumbh Mela preparation, interception and diversion of sewage through channels along the banks towards treatment plants were undertaken. Later the facilities were augmented through JNNURM schemes. A 2011 report of the Municipal Corporation on City Sanitation Plan for Nashik highlighted the presence and spread of green algae, water hyacinth and other species and dead fish in river Godavari is reflective of the inherent pollution risks. It pointed out that the river faced severe risks of pollution concerns from a combination of industrial and domestic sources along with agricultural run-offs. Since then substantial work has been successfully done in certain sectors for arresting pollution. However, the overall city performance is still wanting in adequate monitoring, vigilance and public awareness about industrial, agricultural, domestic waste and solid waste management. News of flouting of pollution norms by contractors responsible for waste disposal and accusations of them getting away with insignificant penalties in regional newspapers are reflective of the same (Lokmat, Nashik 7 November 2020).

Morphology of the Historic Core – Heritage Structures and Precincts and Condition Analysis of the Heritage Fabric and the Intangibles

The historic core is prominently divided into the Nava (New)-pura (suburb) upstream on both the banks and Kazipura, downstream along the southern/western bank. Eight gates in Kazipura and three gates in Nava (also called Maratha) pura, most of which no longer exist except in ruins and names, help in locating the morphological components. The older and more affluent of the settlements occupied elevated grounds of the 'nine hills' while the fertile floodplains which often got flooded were cultivated and later settled by castes and communities of lesser means.

Many of the precincts are known by the traditional occupation of the communities like *tambat* (coppersmith) *ali*, *brahman ali*, etc., *bazaarpeth* (markets) flourish on account of Nashik being an agrarian, industrial and religion-based town and are occupied by communities belonging to trading, jewellery making and other service-oriented occupations.

The historic housing stock consisted largely of richly ornate Hindu (Gujarat) style and/or intricately carved musalman (Delhi) style wadas (mansions); masonary and timber houses ground plus one- or two-storied structures of trading and other communities lined the narrow streets unfit for four-wheeled motorised transport, the only exception being the arterial Agra road (Citation). The aforementioned description of the 1970s has only gotten more chaotic in terms of vehicular stress with further losses of the architectural grain within the historic core in the last four decades. The contemporary generations with better means and seeking houses with amenities, infrastructure and a quieter and more conducive neighborhoods for their children and themselves are shifting to the city peripheries in newer developments. The older generations struggle to maintain their places, renting it for economic reasons or resign to the fait accompli in the demolition and reconstruction of this historic fabric. Many of the traditional occupations too have been rendered obsolete or altered their craft occupations on account of global goods of 'free' market economy.

Temporality of the Kumbh and Its Impact

The 2007 Maha Kumbh Mela was held at Nashik. Facilitating the Kumbh Melas are herculean feats with officials estimating 75 million visitors during the 2007 Mahakumbh of Nashik. Along with the arrangements for the holy dip for thousands of sadhus and lay people, residential arrangements are made in the city on the left bank of the Godavari River. In 2015, the city of Nashik announced that year's Kumbh as harit (green) Kumbh aimed at focusing on the relationship of people and nature. The Kumbh has mixed responses from residents of the historic core – the ones dependent on it for economic reasons welcome it while the people engaged in occupations not directly related to the Kumbh find it an intrusion and during the peak days prefer to move away from their houses for that period. The Kumbh lasts a year with specific peak days, in spite of the municipality's efforts to leave a trail of domestic and ritual related pollution into the river and the city.

Section 6: Towards Cultural Resilience Typology

In this chapter, we learnt through the example of Nashik, how the narrative of resilience in the historic cores needs to acknowledge the specificities of a historic pilgrim town typology especially in the Global South. The emphasis remains more on the chronic stresses which often get overseen in developing the preparedness for acute shocks. The chapter also drew attention to the need of conserving the values and wisdom embedded in morphology of the historic core which otherwise might get erased in our quest for future proofing. In his article, 'Towards a paradigm of Southern Urbanism', Schindler (2017) argues that cities in the Global South constitute a distinctive type of human settlement. He proposes three tendencies that, when taken together, serve as the basis of an emergent paradigm of southern urbanism. Firstly, cities in the south tend to exhibit a persistent disconnect between capital and labour. Secondly, city metabolic configurations are discontinuous, dynamic and contested. Thirdly, political economy is not the overriding context within which urban processes unfold, but rather it is always already co-constituted with the materiality of southern cities. While he presents these three characteristics as non-comprehensive, there are still many more that provide alternatives to 'problematizing southern'. In spite of his intention to 'offer the contours of an emergent paradigm that accounts for the heterogeneity of cities in the South', his arguments preclude the strengths and opportunities inherent due to the cultural aspects in such cities, thus missing a critical component of sustainability and hence resilience. This chapter reinforces Schindler's critique of Brenner and Schmid's concept of planetary urbanisation and aligns with the belief that southern urbanism should be studied without treating these cities as pathological and in need of development interventions. Both these stances are important to construct the resilience narrative of southern urbanism.

Religious urban centres within the state as well as across India have some similar issues related to temporality, difficulties in management of diverse interest of stakeholders, equity in resource allocation and divergent views on sacredness not translating to sanctity of the space and its sacred geography.

Efforts are on to utilise the *Kumbh* to address the issue of disconnect between lifestyles and nature under the call for *Harit Kumbh*. *Kumbh* destinations of Nashik, Ujjain, Prayagraj (erstwhile Allahabad) and Haridwar draw lessons from each other. While the towns of Kolhapur, Pandharpur, Tuljapur, Jejuri, the eight sites of *Ashtavinayakas*, the *Jyotirlingas*, the *Shaktipeethas* within Maharashtra, all need to learn from and inform the resilience narrative of Nashik. Floods affecting the sacred cities of Ujjain along river Kshipra, Kolhapur along river Panchganga and Pandharpur along river Bhima in last two years is a grim reminder of how vulnerable these historic places of religious interest and sacred geographies have become mostly due to anthropogenic activities.

Insensitive infrastructure projects, lack of planned development, inability to harness regional and traditional wisdom, lack of awareness about the value which vernacular living heritage can contribute if conserved need to be attended to. Speculative real estate development is leading to unsympathetic and insensitive demolition of these heritage assets thus incurring irreversible losses of the cultural relics and simultaneously affecting intangible heritage as well.

Acknowledgements

I owe special thanks to Chinmayi Marathe and Pratik Jadhav for preparing the maps specific to this chapter and also to the Nashik Studio KRVIA M. Arch Sem 3 (2019–2021 batch) for the base data used in preparation of these maps.

Note

1 Kumbh is the most important Hindu spiritual gathering in India congregated once every three years, rotating between four major Hindu pilgrim centres of India – Prayag (Allahabad) at confluence of Ganga and Yamuna, Haridwar where Ganga descends from the mountains and commences its journey on the plains, Ujjain on the bank of river Kshipra and Nashik. The Maha (Grand) Kumbh is held every 12th year in each of the four locations. The festival is one of the largest peaceful gatherings in the world and considered as the 'world's largest congregation of religious pilgrims'. The Kumbh Mela has also been inscribed in 2017 on the UNESCO's Representative List of Intangible Cultural Heritage of Humanity (ICH UNESCO).

References

Beel, D. E., Wallace, C. D., Webster, G., Nguyen, H., Tait, E., Macleod, M., & Mellish, C. (2017). Cultural resilience: The production of rural community heritage, digital archives and the role of volunteers. *Journal of Rural Studies*, 54, 459–468. ISSN 0743-0167. Retrieved November 25, 2020, from https://doi.org/10.1016/j.jrurstud.2015.05.002

Bhardwaj, S. M. (1994). The concept of sacred cities in Asia with special reference to India. In A. K. Dutt, F. J. Costa, S. Aggarwal, & A. G. Noble (Eds.), *The Asian City: Processes of Development, Characteristics and Planning. The GeoJournal Library*, vol. 30. Dordrecht: Springer. https://doi.org/10.1007/978-94-011-1002-0_5

Binder, C., Wyss, R., & Massaro, E. (Eds.). (2020). *Sustainability Assessment of Urban Systems*. Cambridge: Cambridge University Press. https://doi.org/10.1017/9781108574334

Camuffo, D. (2019). Climate change, human factor, and risk assessment. *Microclimate for cultural heritage,* 303–340.

Clarke, N. J., Kuipers, M. C., & Roos, J. (2019). Cultural resilience and the Smart and sustainable city: Exploring changing concepts on built heritage and urban redevelopment. *Smart and Sustainable Built Environment*, 9(2), 144–155. https://doi.org/10.1108/SASBE-09-2017-0041

Daskon, C. (2010). Cultural resilience – The roles of cultural traditions in sustaining rural livelihoods: A case study from rural Kandyan villages in Central Sri Lanka. *Sustainability*, 2. Retrieved November 25, 2020, from https://doi.org/10.3390/su2041080.

Hawkes, J. (2001). *The Fourth Pillar of Sustainability: Culture's Essential Role in Public Planning*. Retrieved November 25, 2020, from www.researchgate.net/publication/200029531_The_Fourth_Pillar_of_Sustainability_Culture's_essential_role_in_public_planning/citation/download

Holtorf, C. (2018). Embracing change: How cultural resilience is increased through cultural heritage. *World Archaeology*, 50(4), 639–650. https://doi.org/10.1080/00438243.2018.1510340.

http://nashikcorporation.in/public/upload/download/5_Ch2_Study_Area_Nov20_2.pdf. Retrieved November 25, 2020, from http://nashikcorporation.in/public/upload/download/5_Ch2_Study_Area_Nov20_2.pdf.

http://nashikcorporation.in/public/upload/download/10_0_Nashik%20Draft%20CSP_Executive%20summary.pdf. Retrieved November 25, 2020, from http://nashikcorporation.in/public/upload/download/10_0_Nashik%20Draft%20CSP_Executive%20summary.pdf

https://cultural.maharashtra.gov.in/english/gazetteer/Nasik/005%20History/001%20AncientPeriod.htm. Retrieved November 25, 2020, from https://cultural.maharashtra.gov.in/english/gazetteer/Nasik/005%20History/001%20AncientPeriod.htm.

Jigyasu, R. 2013. *Heritage and Resilience: Issues and Opportunities for Reducing Disaster Risks. Background paper*. Retrieved May 24, 2020, from http://icorp.icomos.org/wp-content/uploads/2017/10/Heritage_and_Resilience_Report_for_UNISDR_2013.pdf

Maser, C. (1992). *Global Imperative: Harmonizing Culture and Nature*. Stillpoint Pub, Michigan. ISBN 10: 0913299901; ISBN 13: 9780913299906.

Schindler, S. (2017) Towards a paradigm of Southern urbanism. *City*, 21(1), 47–64. https://doi.org/10.1080/13604813.2016.1263494

Shinde, K. A. (2011). Placing communitas: Spatiality and ritual performances in Indian religious tourism. *Tourism Preliminary Communication*, 59(3), 335–352. UDC: 338.48–6:2(540). Retrieved November 25, 2020

Shinde, K. A. (2016). Planning for urbanization in religious tourism destinations: Insights from Shirdi, India. *Planning Practice & Research*. Routledge. https://doi.org/10.1080/02697459.2016.1198197

4

URBAN WATER RESILIENCE

Jamshid Bhiwandiwalla

Section 1: Traditional Water Systems in Historic Cities

Historically, Indian cities have survived by managing their surface waters, much before they could master tapping of the ground and piped waters. Urban settlements over the ages have showcased unique water harvesting systems and techniques, thereby respecting the watershed areas that feed these systems as well as their position in the urban realm, thereby building strong sociocultural links with the community. Elaborate descriptions of policies and strategies with regard to water sufficiency in the kingdoms have been found in Kautilya's *Arthashastra*. Prehistoric *Dholavira* in the arid dessert showcased the ideal way that harvested surface water for survival and its model is a threshold for urban planning. The Western ideas of having temples atop the acropolis were replaced by the great tank and granary in most of our Harrapan cities. Dynasties in North India exhibited unique canal systems to tap glacial waters of the trans-Himalayan regions, whereas as South India saw the building of check dams to hold rain waters along with elaborate temple tanks (*kunds*) to serve the adjoining settlements. This resulted in associating water as a resource with a sacred status. Temple tanks in the renowned Khajuraho precincts cease to exist today, whereas those similar in the temple town of Madurai in South India exist and serve for daily rituals of devotees. Rajput queens in West India built elaborate *Baolis, Vavs, Jhalaras* (stepwells) as social amenities for women. The everyday practices of fetching water came to be celebrated with opportunities for social interactions, episodic festivals and rituals around them.

Stepwells, most of them introvert complexes of multistoried galleries of intricately carved stone colonnades, serve as resting and social areas for women, away from the male gaze. These complexes today apart from the historicity remind us of the strong sociocultural connect that was in place

DOI: 10.4324/9781003098461-4

Figure 4.1 Elaborate and shaded stepwells along western India served as a recluse and social amenities for women.

Source: Unsplash

with water infrastructure as a public realm. Carefully planned, these structures not only tap on the groundwater at one end but also hold onto the entire surface water runoff of the region, thereby submerging galleries and limiting its use during heavy showers.

In medieval India, the Mughal rulers used their Persian influence to tap underground streams in arid regions and adopted the *Qanat* canal systems. Streams tapped in the high hills and water channelled to the city to irrigate their *Char-baugs through the Nehars* exist till date at Burhanpur in Madhya Pradesh as well as at Aurangabad in Maharashtra. These *Nehars* were punctuated with air shafts and access to the subterranean using the distinctive masonry vaulted galleries. Other than having water tanks along the high ground of the Mughal forts to serve the royalty, remnants of Aqueducts and the Persian wheel can still be found at the threshold of highly placed Mughal fortifications supplying water to the *hammams and serais* serving the traders and soldiers stationed within the ramparts. It is evident from this discussion that there persisted an intimate connection between water resources and its people and their sociocultural practices that were revered, cherished and maintained for centuries.

Figure 4.2 Mughal architecture has always associated with Persian concepts of *Char-baug* and Qanat water systems.

Source: Unsplash

Section 2: The Disconnect Between People, Culture and Water

The legacy of piped water has been handed over to us from the British colonial rule. It was the British colonial rulers in India who considered open surface water catchments as well as the activities around them as highly unhygienic. English engineers brought water through the mechanism of pumping from untouched areas afar irrespective of the geology, topography or hydrology of these geographies. It soon became a marker of progress as piped water reached rural hinterlands. Water available at the doorstep was a luxury and came to be symbolised as modern living that soon became a norm. After her independence in 1947, successive governments in postcolonial India aimed to send piped water to every household and the latest target set is by the year 2024. Although a worthy cause, this simultaneously led to ordinary people losing their connect with water harvesting as well as management of these historical systems that are often community-driven. Cities were planned for water supply but not for wastewater, thereby polluting our primary sources of water like rivers turning them into drains or *nullahs*. In the current times, cities rely on catchment areas over hundreds of kilometres away, and all efforts towards supplying piped water do not address the issues of water scarcity with growing demands from an ever-expanding population and rapid urbanisation. On the one hand, the demand for 2% of the household's domestic water supply through piped infrastructure always falls short as there is no judicious use of portable water. We have taken for granted water as a resource, thereby leading to extensive wastage. Existing infrastructure laid over the centuries in historic second-tier cities of India holding surface water is put to complete disuse due to extensive pollution and neglect.

India's water scarcity was brought to the fore with extreme water shortages in major cities, particularly in Chennai. Experts, environmentalists, and NGOs have been warning about India's impending water crisis for a long time. They failed to grab the attention of all until the taps went dry in big cities. The NITI Aayog released a report in June 2019 in which it acknowledged that the country was suffering from the worst water crisis in history and about 600 million people or about 45% of the Indian population suffers from high to severe water stress (https://niti.gov.in/content/composite-water-management-index, 2018). Further, the report predicted that 21 Indian cities would run out of groundwater by 2020, nearly 40% of the population will have absolutely no access to drinking water by 2030, and 6% of India's GDP would be lost by 2050 due to water crisis (Chakrabarty, 2019).

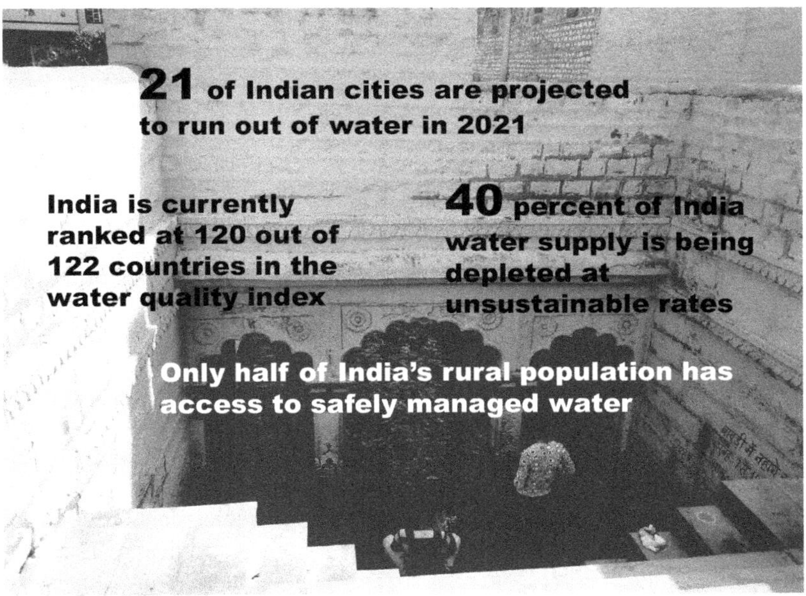

Figure 4.3 India's water crises.

Graphics and image source: Author

India's water woes and scarcity stemmed primarily from its over-dependence on groundwater and piped water, leaving existing age-old surface water infrastructure to disuse. Being the world's largest user of groundwater, not only are our cities most groundwater dependent but also is our irrigation, thereby extracting 250 cubic km of groundwater annually and has 20 million wells and tube wells (www.hindustantimes.com/analysis/india-must-not-look-at-its-water-crisis-in-isolation, 2019). India, on the other hand, receives extensive rainfall, compounded by erratic climate change; however, cities have paved over wetlands and encroached over lakes, thereby not helping the process of recharging the water table. Poor land use and watershed management lead to flooding and causing economic devastation year after year as rains could not percolate into aquifers under the city through appropriate surface water management. *Those who have understood this phenomenon for long understand that these are a manifestation of the same malaise – decades of mismanagement of our land, water and ecosystems due to outdated and misinformed policies.*

Section 3: Water Resilience and Southern Urbanism

Resilience is commonly understood as the ability of systems (social or bio-physical) to withstand or cope with stressors while continuing to maintain key functions or structures (Folke et al., 2016). In the context of water systems, various definitions of resilience are in use: for example, *engineering resilience, measuring the attributes of engineered water systems and their ability to bounce back from disruptions; ecological resilience, focusing on the capacity of eco-hydrological systems to cope with stress; or community resilience, focusing on the ability of society to cope with water stressors or risks* (Rodina Lucy, 2019).

A focus on strengthening resilience can protect development gains and ensure people have the resources and capacities to better reduce, prevent, anticipate, absorb and adapt to a range of shocks, stresses, risks and uncertainties (Aditya Bahadur, 2015). Building resilience in water systems of what it constitutes and how it may be achieved is still not well understood, due to want of theoretical or practical guidance, considering that water systems are '*highly complex, highly fragmented, and typically compartmentalized across disconnected sectors*' (Rodina Lucy, 2019). Further in the resilience scholarship, outside the ongoing epistemological and methodological debates, there is an understanding on the type of water management practices and governance arrangements that can increase resilience in water systems. '*A growing body of work is increasingly arguing for more integrative ways of moving toward water resilience to encompass the various aspects of water systems (social, infrastructural, or ecological)*' (Rodina Lucy, 2019).

In the context of water resource management and governance, water resilience is increasingly used in relation to social systems and, particularly, is typically defined as the political, organisational and administrative processes whereby communities articulate their interests. Their inputs are subsequently absorbed based on the decisions made and implemented. *Decision-makers are held accountable for the development and management of water resources and delivery of water services* (Rodina Lucy, 2019).

Further, historically worldwide engineered water systems have been inflexible and slow to adapt to change as they are embedded in infrastructural legacies and design paradigms (White, 2010). It is best to have a collaborative approach incorporating people, their social and community structures and cultural relations with the natural ecology.

However, restoring healthy ecosystems is considered the most important strategy in the Global South for dealing with uncertainty and increasing resilience in water systems with the ability to quickly respond to change. Green infrastructure via soft solutions, as opposed to grey infrastructure is considered as the most prominent approach to flood resilience followed by integrated land use and water planning with diversity and response options. *Resilient, adaptive infrastructure such wetlands cannot be built. They grow slowly and extensively building up relationships in steps and bounds, integrating into surrounding systems, flows and entities that evolve over time until they are part of the essential and become natural and invisible* (Carlisle, 2013). One such city that historically exhibits a system in green infrastructure and productive landscape operating both at a regional and an urban scale is the well-known East Wetlands of Kolkata.

Based on the banks of the river Hoogly, Kolkata is prone to flood, considering the river level is higher than the existent colonial city. Further, the east of the city has extensive wetlands through which the entire waste of the city flows before it enters the river basin that plays a pivotal role in detoxifying waste, filtering water and providing a much-needed natural flood barrier. Most unique to southern urbanism is the fact that these wetlands are regulated by the local communities creating a sociocultural connect, generating employment and food through fish and vegetables. The produce here feeds the entire city and also assimilates the city's waste and cleanse the forming a self-sufficient, sustainable ecosystem.

Globally, wetlands have been under threat, and according to one estimate, the world has lost half its wetlands in the last century with India having lost 38% of its share in the last decade. Thankfully, these sensitive ecosystems, 37 of these wetland sites in India, have been notified for protection under the international treaty signed at Ramsar, Iran, in 1971. 'While wetlands in developed countries are cherished for their biodiversity and aesthetic value, wetlands in India are linked to survival', says Parineeta Dandekar, Associate

Figure 4.4 Kolkata wetlands being lost to encroaching urbanisation.
Source: Unsplash

Coordinator at South Asia Network on Dams, Rivers and People (SANDRP). Just as the East Kolkata wetlands, in other sites such as the Vembanad wetlands in Kerala, hectares of paddy cultivation are well irrigated, and the Bhopal wetlands (lakes) provide drinking water to the entire city (https://economictimes.indiatimes.com/news/politics-and-nation/vanishing-wetlands-indiscriminate-development-poor-regulation-are-wrecking-a-critical-piece-of-indias-ecology/articleshow/51473398.cms, 2016).

Ritesh Kumar, Conservation Program Manager, Wetlands International – South Asia, an advocacy and research outfit, who believes that there is a vast difference in approach and investments dedicated to the upkeep of wetlands, contends:

> Across India, ranging from big cities to small towns and even deep into the countryside, wetlands are often perceived as wastelands and easy pickings when space is required for real estate, infrastructure and other forms of commercial development. While this has been the norm for decades now, recent floods have now put the

spotlight on this ignored resource. Common lands that were meant for everyone are being exploited by a few.

He adds, 'Not all of them have received the same attention . . . many have no management plan at all'. Further, he asserts, 'As the weather has become more unpredictable and we deal more with the effects of climate change, the role of these wetlands as a buffer has only become more important'. Government, corporate and stakeholder communities will soon need to connect the dots between disaster, development and water management (https://economictimes.indiatimes.com/news/politics-and-nation/vanishing-wetlands-indiscriminate-development-poor-regulation-are-wrecking-a-critical-piece-of-indias-ecology/articleshow/51473398.cms, 2016).

Section 4: Case of Bhopal Wetlands

More recently known for its Union Carbide Gas tragedy, Bhopal, also known as the city of lakes was once placed among the greenest cities of India, is fast losing its tag of a green city. According to a study by researchers from IISc Bangalore, Bhopal's vegetative cover of 92% in 1977 has been reduced to 21% in 2014. The study predicts degradation of the same to 11% by 2018 and just 4% by 2030 if the city administration continues to work on the same policies (Bharath Aithal, 2016). So have the lakes shrunk too appreciably though they are Ramsar protected sites? The Upper Lake has reduced from its initial 30 sq. km to 8 sq. km as of now. The Lower Lake also got reduced from 8 sq. km in the beginning to 2 sq. km in 2009. The total length of the lake was 38 km, but it reduced to 5 km in 2009 (www.rainwaterharvesting.org/bhoj_lake/bhoj_lake.htm, 2009). However, historically Bhopal boasts of the first-ever recorded attempts by Raja Bhoja of creating the largest human-made lake around at 1010 CE. Many are unaware that there existed a much larger Sagar Taul (*sagar* means the sea and *taul* means lake) or Bhima Kund spread over 650 sq. km almost 47 times of the former by cleverly exploiting the terrain and just building three infill dams in between relatively small gaps formed by two hills (Dass, 2011). The Sagar Taul was destroyed in 1334 by Hoshang Shah as it took an army for three months to destroy the dams and water flowed for three years, thereby altering the climate of the Malwa region as recorded. Many settlements down the Betwa River were destroyed due to flooding and there was a lake bed that was not habitable for the next 30 years; the massive stone blocks of the dam strewn along the valley is a grim reminder of the once-prosperous Bhojpur city along its banks.

Today, most of Bhopal city sits around the Chota Talab, a lake extend in 1794 CE by building a PulPukta, a dam built across the Ban Ganga and

Patra valley that took the overflow of the *Bara Talab*. However, the 1700s saw the building of the citadel of Fatehgarh on the highest plateau to the north of the Bara Talab. This fortified Bhopal city, also known as Sher e Khas, did boast of *hammams* and lavish *serais* as recorded by travellers. The 1800s saw the reign of Sikandar Jehan Begum, who commissioned English engineer David Cook to develop the lakefront and plan the waterworks system. This is the era of major construction projects like the Jami Masjid and the Moti Mahal along with the construction of 72 wells of which only ten exist in poor condition like the one at Bara Baug were three-storied stepwell, which served public gardens the city was well known for. Further, 1870 saw the ambitious development of the suburb of Shahjehanabad to the north of Bhopal for the elite *nawabs* sustained by another set of three terraced cascading lakes, the Motia Talab, Noormahal Talab and the Munshi Hussaini Talab dependent on seasonal rains and surface runoffs. Further, the remnants of an aqueduct at Jehangirabad are one of the few amenities that showcase the rich water infrastructure laid to support the city for 1,000 years interwoven with the cultural fabric of the city (Dass, 2011).

However, over the past few years due to the indisposition and lack of initiation from the local authorities, there have been a continued degradation

Figure 4.5 Bhopal, city of lakes.

Source: Author

of the environment of the city. The last master plan that was approved for the city was in 2005, while the later editions did not earn approval from the people and executive alike, and therefore, the city is progressing without one for the past 11 years (Mrunmayi Wadwekar, 2018).

The absence of a master plan has meant that the city has expanded without due consideration of its land uses, planning guidelines and resources. Watersheds have been encroached upon by housing developments and service infrastructure and so have lakefronts by way of informal settlements, thereby leaving sewage into the lakes. Water quality of the Upper Lake which could be directly used till a few years back now needs at least primary treatment before consumption. Among the other lakes too, problems with the quality of water are noticed, several of them in various stages of eutrophication. Also, according to a study by an independent newspaper in Bhopal, among the 31 registered lakes in and around Bhopal, only 21 exist as of 2016 while 11 have been permanently lost (TeamDB, 2016).

However, unlike as stated previously, the southern paradigm has always exhibited a strong connection between the water resources and inhabitant residents. Fishing and other occupational-based communities have always lived symbiotically with the wetlands, thereby in many a way protecting its edges through community participation. Recent mapping as well as studies along the edges of the Upper and Lower lakes of Bhopal by the Urban Design and Conservation students of semester II, KRVIA have shown the disconnect between the communities and the water resources, considering there are several polluting activities by informal settlements as well as unplanned gated communities along fringes. Irrigation lands in the vicinity of the city are being bought by land speculators, and that has led to increasing prices as well as the loss of agricultural land and associated livelihoods. This has directly contributed to the degradation of the environmental quality along edges of the wetlands, by way of pollution as well as flooding.

To further assess the impacts of urbanisation, remotely accessed data were used to study temporal land use and land cover change. Four types of land use categories were identified for analysis, namely: agriculture, built-up, barren and others. Later, a stream order map was generated, which when superimposed with the built-up category revealed the pattern of watershed modification. The same raster images were then used for FRAGSTATS analysis that revealed the changes to the vegetation and landscape of the areas. The analysis showed that agricultural land was reduced by 45% while built-up land was increased by almost 260% from 1972 to 2016. Streams of the first and second order essential for carrying the storm flushes are blocked or modified, thereby modifying the watershed. The overall landscape has become fragmented over the years and that there is a loss of biodiversity (Mrunmayi Wadwekar, 2018).

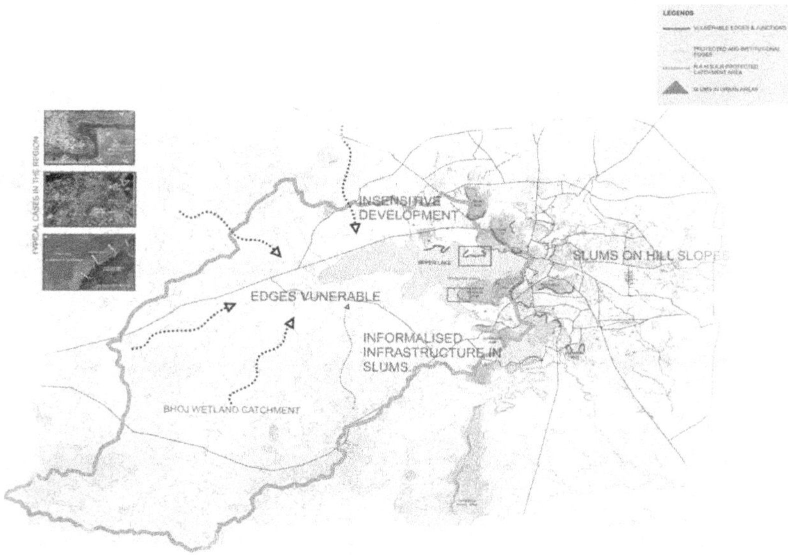

Figure 4.6 Extent of the Bhoj Wetland as notified under the Ramsar Convention, location and size of Bhopal's upper and lower lakes and its vulnerable edges with regard to the city.

Source: KRVIA Masters Program Sem II 2019

Groundwater in the city is also facing problems, as vegetation in the peripheral areas has reduced, leading to depletion and degradation. The localities in and around Union Carbide are affected due to severe contamination in the soil and groundwater. In other areas of the city, pollution of groundwater sources due to improper disposal of wastewater is being identified through scientific studies. A study on impacts of urbanisation on surface and subsurface flows conducted for Bhopal city concluded that because of the natural topography and geological characteristics, storm events had manifested themselves in an altogether different way and that urbanisation not only magnified peak discharges but also created new peak discharges. Therefore, the floods of any magnitude now occur more frequently. Increasing urban flooding events in recent years have been a testament to this study (Mrunmayi Wadwekar, 2018).

The usefulness of history for the understanding of the present and future is generally agreed, and it is quite common for non-specialists to argue that those who do not learn from history are doomed to repeat it. Temporality

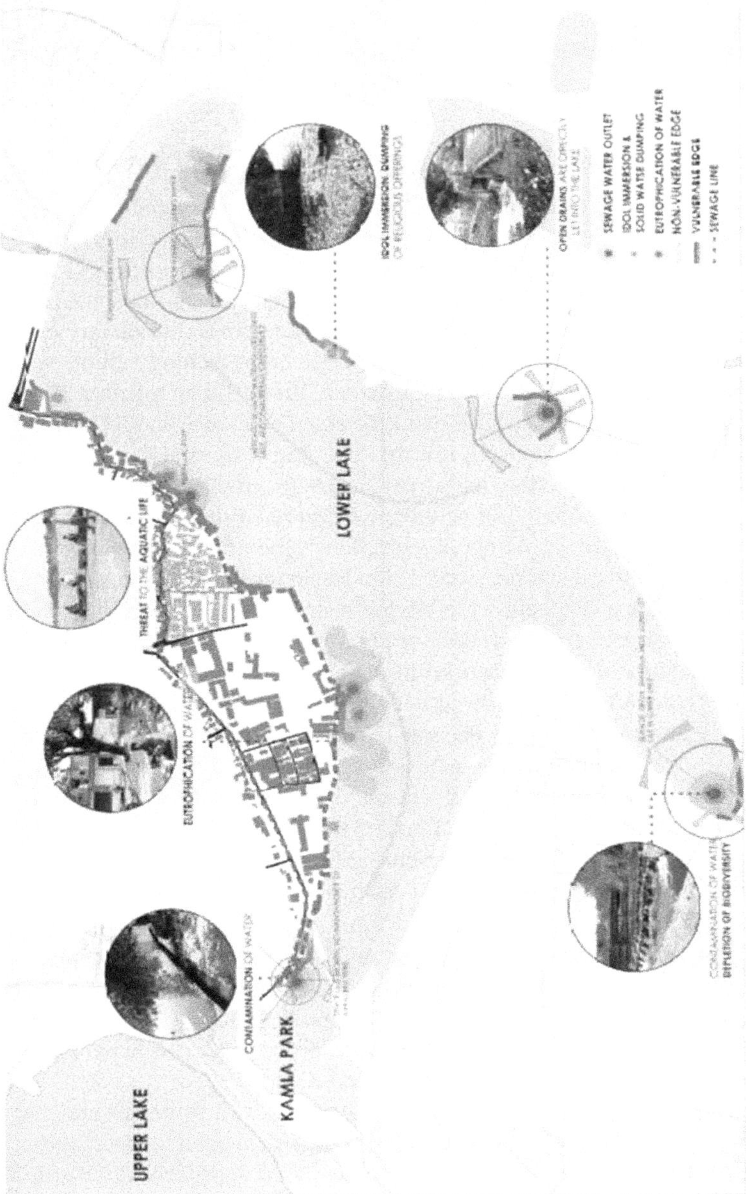

Figure 4.7 Mapping exercise of Bhopal's lower lakes and its vulnerable edges along the city.

Source: KRVIA Masters Program Sem II 2019

in terms of the relationship between past, present and future, the notion of temporal scales, concerns the time frames in which we think about urban resilience. If resilience discourses are tied to debates about sustainability and climate change, questions of duration need to be posed with a long-term view of the future (Brantz Dorothee, 2020).

Section 5: Resilient Strategies for the Global South

'Resilience thinking has been applied to everything from human development to systems engineering, and this is one of the reasons that critics believe the terminology has become hopelessly vague' (Brantz Dorothee, 2020). From its etymological origins in the 1620s to its present-day use, the term resilience has been framed in numerous ways and across disciplinary contexts from philosophy to engineering, planning and psychology all the way to ecology and the social sciences (Alexander, 2013). Taken together, these discourses provide a genealogical narrative about resilience and its intrinsic norms and values. The first step towards building resilience is identifying stakeholders and constructing narratives across geographies and scales that are impacted or impactful. The paradigms of southern urbanism show that these stakeholders are incredibly diverse, these other than being human are physical, sociocultural and financial, not to forget most of the time affecting the flora and fauna. However, identifying stakeholders and mapping their narratives through exercises as seen in Figure 4.6 in the case of lake edges of Bhopal legitimise their intentions through the stories they tell. No wonder urban critics and community activists have a different take on resilience measures than planners and bureaucrats, as those not affected by crises have a different view about measures to be adopted from behind a desk.

Recently, there has been a significant discursive reframing of urban development efforts away from notions of sustainability towards practices of resilience. This narrative shift can be clearly traced in the literature starting in the early 2000s. A 'narrative of resilience' rather than urban sustainability appears to be the new urban paradigm, and this narrative shift needs to be critically evaluated (Sudmeier-Rieux, 2014). In our view, disruptions to everyday life – from floods to uneven access to water – must be addressed both in terms of their immediate causes and effects and in terms of their long-term drivers and desired outcomes. This has been addressed by the triforked strategy of resilience given in Figure 4.8.

While there exists a flexibility on the aforementioned strategies adopted which clearly builds a logic for the term itself based on the relevant context, the first two methods however are top-down–based approaches to disruptions with key roles played by planners, architects, engineers and bureaucrats. The last of the strategies largely associated with the grassroots level

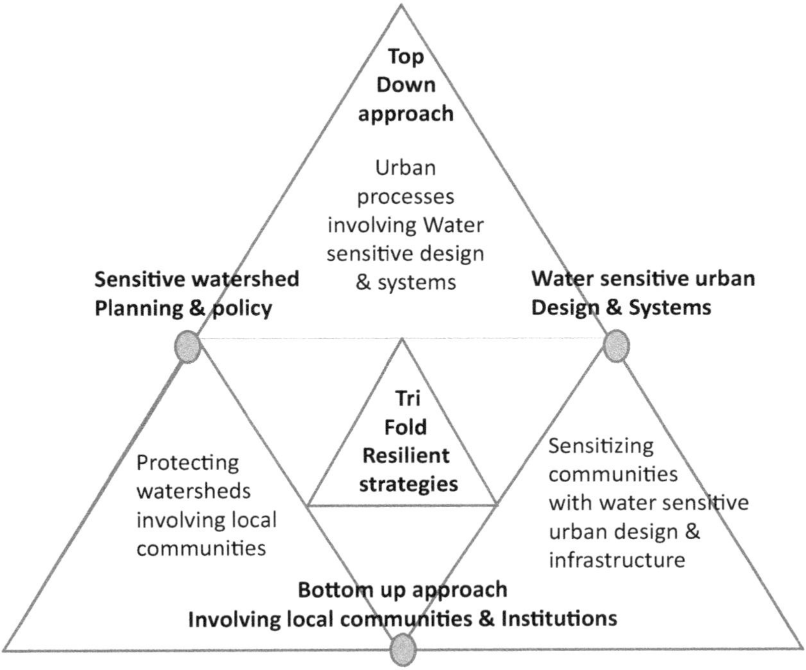

Figure 4.8 Tri-forked resilient strategies.

Source: Author

by way of involvement with local communities and institutions as a bottom-top approach found to be more effective in the southern paradigm. Using resilience in its contextual, vernacular and everyday sense creates a space for negotiation between different sets of strategies. It opens up the possibility for new common understandings to emerge, level thereby resolving ideological tensions within resilience discourse and help to generate a framework that would make resilience both more concrete and more conceptually robust (Brantz Dorothee, 2020).

Sensitive Watershed Planning and Policies

Resilience has been applied to a range of issues and at a variety of scales – from global financial and ecological systems to human development – but cities have become a particular object for resilience approaches (Chandler,

2016). Second-tier cities that are at the cusp of development thereby showcasing unprecedented growth need to learn from the planning mistakes, leading to frequent disruptions and disasters of our metropolises. It has been found that poor understanding of the existing watersheds and associated land uses in planning procedures are the foremost concerns in resilience building. Advocates of resilient cities believe that planning of watersheds as a powerful tool serves to be the most relevant policy agenda in safeguarding our cities, and this does need to be critically evaluated.

In their edited volume on resilient planning, Eraydin and Taşan-Kok argue that this tension between practice-oriented disciplines and critical urbanists is itself a product of neoliberalism. Planning, they argue, has since the 1970s 'become increasingly market-oriented and entrepreneurial. . . . All around the world, urban development has become increasingly fragmented . . . with opportunity-led planning practices taking root everywhere in reaction to rapid and complex change' (Eraydın, 2013). In their own calls for a shift towards a 'resilient planning' paradigm, Eraydin and Taşan-Kok argue 'that architects and planners have increasingly been forced to design and plan for the short and medium-term, to package and sell plans to stakeholders who are committed to market principles, and this is an important point' (Brantz Dorothee, 2020).

Cities such as Chennai, Bangalore and recently Hyderabad have faced disasters such as floods due to their systematic disruption of existing and natural watersheds. At the same time, water sustenance through natural and artificial lakes in hot and arid cities like Jodhpur and Udaipur have survived over the ages as their watersheds have been conserved as well as restored. Jodhpurs intrinsic water system interwoven with the city fabric comprising *jhalaras, baolis* and stepwells have survived only because the primary watersheds of the source Ranisar and Padamsar water tanks have been well conserved through the Rao Jodha nature sanctuary.

Concerning water governance and policy, several authors have argued that polycentric governance enhances resilience shifting from state-led often hierarchical, top-down governance of water to include community-based organisations and civil society that share authority and responsibilities in water management. Such 'independent but coordinated centres of authority' at the community scale at which they occur are theorised as better able to respond to water resilience (Rodina Lucy, 2019). *Resilience scholars also argue that to enhance the resilience, as well as the transformative capacity, of urban water systems, there is a need for a mix of centralized and decentralized governance forms, as well as a mix of formal and informal institutions* (Rodina Lucy, 2019). Such is the case of our national water policy whereby three independent national programmes such as the

Ministry of Water Resources, River Development and Rejuvenation and Ministry of Drinking water and Sanitation are now merged into one centralised programme, the National Jal Jeevan Mission by the ruling government to address concerns related to water issues in a coordinated manner, considering they are so interrelated. Considering the concentrated efforts, the momentum on the nationwide sanitation programmes has achieved unprecedented success. Similarly, the Ganga rejuvenation and nationwide river connect project is also looking at addressing crises of both water scarcity and regional flooding.

While the proposition is to think of water systems in a more integrated and adaptive way (Biswas, 2009), the hypothesis is presented as being able to solve complex problems through decentralised approaches by involving a wide diversity of stakeholders through strategic collaboration and, therefore, a wider diversity of knowledge to learn from.

Water-Sensitive Urban Design and Systems

Rapid urbanisation modifies the topography by creating impervious surfaces, thereby drastically altering the hydrological cycle, causing either water shortages or flooding. Instances from history have shown Indian cities to have had manifests spatially to accommodate water as a resource as rivers, streams, lakes, stepwells and tanks. These have also contributed to the sociocultural and, at times, religious context, thereby contributing to a flexible as well as a diverse and holistic approach to Urban Design. Today, the need is to revive these inbuilt neglected systems which help foster resilience, by adopting multiple and diverse sources of water (e.g. surface water, groundwater), using rainwater harvesting to enhance surface and groundwater sources, recycling water for gardening and toilet flushing, which in turn requires new infrastructures and regulations to balance public good and other concerns. Hence, an underlying ecological framework that forms the backbone of the city with water-sensitive urban design (WSUD) paradigm as well as policies favouring water governance that are associated with the various dimensions of the water sector as well as social systems, thereby aiming towards a more eco-centric, holistic and sustainable approach to creating 'water resilience'.

The literature on water resilience puts forward a large number of complex and diverse propositions that tend to draw on many different aspects, scales, characteristics, or types of water systems, be they biophysical or social (Rodina Lucy, 2019). Flexibility and diversity of response options in the context of natural resource management have been proposed as key

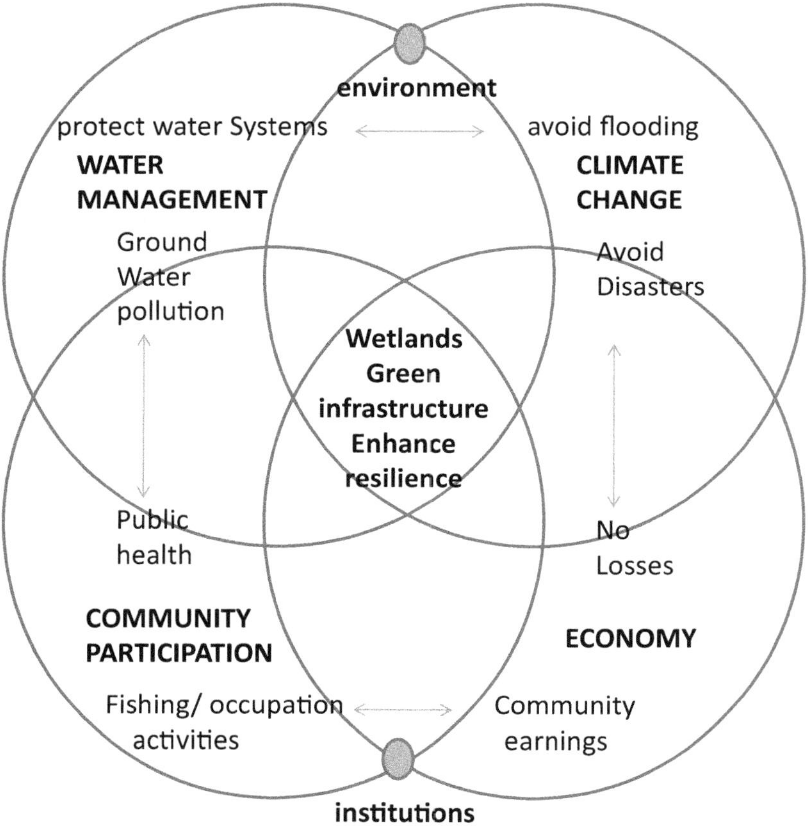

Figure 4.9 Resilience framework.

Source: Author

resilience enhancing strategies because they allow the social system to adaptively respond to change (Schluter, 2007).

In the case of Bhopal and other Ramsar sites, though the holistic approach to water resilience is seen as the building of green infrastructure, as a system via wetlands, such transformations need to increase the ability of water systems to deal with hydrologic uncertainty and unpredictability such as climate change and pollution, and to increase connectivity across scales – in other words, to increase water resilience to various complex and emerging stressors.

Involving Local Communities and Institutions

During disruptions and disasters, it is proven that the poor, marginalised and weaker section of the society, as well as a gender, are the hardest hit. This is particularly true in the Global South where disasters affect race and class dynamics of an individual and in return affect financial and material sourcing of aid. Similarly, it is well understood that not all challenges can be resolved by the state and that a bottom-up approach to disruptions in all its diversity can be addressed at the grassroots level by engaging vulnerable stakeholders and local communities, thereby giving them a voice to respond to disasters. Local communities create symbiotic relationships with the environment for their food production and financial sustenance; however, there are instances when their relationships are at odds with resilient strategies.

Stakeholders like fisher folk/agriculturist and potters/idol makers who reside beside water edges have a symbiotic relationship by way of sociocultural and financial sustenance with the resources; on the other hand, marginalised communities in the form of informal settlements or urban development along the fringes pollute holding ponds/wetlands by dispersing waste into the same. The Bhopal lakefront exhibits both scenarios, thereby bringing vulnerable stakeholders to the fore along with the co-creative approaches to effective crafting to resilience strategies. Another bottom-up approach to resilience building is empowering local institutions in safeguarding water resources like in the case of Jodhpur where all freshwater wells are protected by religious institutions. Jodhpur also has the distinct privilege by way of a case study of urban rejuvenation by a local NGO, the JDH initiative in a one of a kind Toorji ka Jhalara precinct revival, thereby creating a cultural and financial resilience to an earlier decaying historic core area.

As resilience discourses have gained in popularity, though, they have also generated opposition. Some critics argue that resilience is part of a larger neoliberal project that leverages real or perceived crises to justify policy agenda that would otherwise be unpalatable to the public and the international community (Cretney, 2014). Most obviously, critics are concerned about the way that resilience is used to push non-governmental solutions to challenges that have typically been the responsibility of the state.

The Way Forward

Cities all over the world are vulnerable to disruptions, and many of them have faced disasters; however, historically cities have never been abandoned due to several reasons, one being that resilience is the DNA of the urban along with urban development accompanied by property rights. It can also be well understood that cities are a repository of shared memory and civic pride and recovery from any disaster are of great symbolic significance. Hence, the new buzzword of urban development is the narrative of resilience which needs to be critically evaluated so that it just does not serve as a planning tool but also functions as a robust policy agenda.

Lastly, in the case of urban water resilience in the Global South, the sectoral fragmentation and disciplinary legacies in the water sector are very persistent and difficult to overcome; however, a robust framework that overcomes the ideological tensions to build a concrete response as a 'resilient responses' versus 'resilience building' should be different by way of context as well as systems adopted, thereby building a resilient ecosystem.

Box 4.1 Resilience practices in India

Water conservation and revival: Story of '*Johads*' in Rajasthan

In Rajasthan, rainwater conservation was done through 'Johads', earthen check dams made in large crescent shapes with rocks and earth, a practice followed since 1500 BC. This ancient method was revived by Rajendra Singh, renowned conservationist from Rajasthan and well known as 'Waterman of India'. During his visit to Gopalpura village in Rajasthan in 1985, he discovered that villagers desperately needed water more than healthcare due to dried wells and rivers. So, through his founded NGO, Tarun Bharat Sangh (TBS), he started to dig and built a Johad at Gopalpura, with villagers. It took 7 months and almost 3–4 people per household to build a 20-foot deep and 1,400-foot long 'johad'.

By 1996, Gopalpurans had built nine johads, covering 2,381 acres and holding 162 million gallons of water. Underground water had risen from an average level of 45 feet below the surface to only 22 feet, and all the wells had water. Fuel prices reduced (less pumping required) by 75% and the area of wheat fields jumped from 33 to 108 hectares. Spectacular success of reviving wells and collecting rainwater had profound ripple effects. Over next 25 years, Rajendra Singh and his NGO, TBS played a catalysing role in building of 8,600 johads in 1,058 villages spread over 6,500 sq. km in nine districts of Rajasthan, including Alwar, Jaipur, Jodhpur, Jaisalmer, Karauli, Pali, Sawai Madhopur and Tonk. More recently, TBS has consolidated its work in four additional districts of Alwar, Karauli and Sawai Madhopur.

- As a ripple effect of these low-tech check dams, the groundwater level rose from about 100 m to about 13 m. The green cover in the region also increased from 7% to 40% and land under cultivation increased by 30% to 150%, making available more food, fodder and income for its villagers, and cattle wealth improved for traditionally pastoral communities.

- The new rivers and watering holes provided sustenance for wild animals. Bhaonta-Kolyala reforested 1,500 acres of the neighbouring hills and declared it a people's wildlife sanctuary.
- Once known for dead rivers, the region has recorded documentation of recovery of the Arvari River and its fishery, so much so that there was a political struggle over fishing rights.
- Social benefits include the rise in confidence and pride that comes from community self-governance.

Important quotes and reads:

1 **'Water Johads: A Low-Tech Alternative to Mega-Dams in India' (Vansintjan, Aaron, NoTechMagazine.com)**: *The key innovation with the johads is that rather than relying on engineering expertise or governmental action, villagers have constructed the johads themselves through traditional methods and community participation.*

2 **'Water Warriors: Rainwater Harvesting to Replenish Underground Water' (Suttari et al., 2005):**
The rebirth of rainwater harvesting set loose a cascade of constructive forces, in Gopalpura and beyond. The effects ping-ponged from ecosystem to social system and back, and the momentum got stronger, as both systems began to heal themselves. The virtuous cycles were mirror images of the vicious cycles that preceded them:

- ***Success breeds success.*** *As their wells filled with water, villagers were encouraged to build more johads, bringing even more wells back to life. They revived the traditional gram sabha village council, which planned, built and maintained more johads.*
- ***Trees and underground water.*** *The rise in the water table encouraged villagers to plant trees. The trees and other vegetation protected the soil, reducing erosion and siltation of johads, allowing more rainwater to seep underground and raise the water table even more.*
- ***Community manpower.*** *As there was enough water for a dry season crop, young men moved back from cities to the village, providing more labour to construct and maintain johads.*

3 **'Ingredients for Success – Rainwater Harvesting in Rajasthan' – Gerry Marten:**

1. **Shared community awareness and commitment:** Working only for food, a number of Golpapura villagers joined the team to restore the first johad. The following year, a larger dam was restored by the residents with an estimated 10,000 person-days of labour. Traditional participatory village councils (Gram Sabha), which featured representation from every family and reached decisions by consensus, were revived to manage dam construction. The Gram Sabha also initiated community reforestation projects. These cooperative efforts strengthened village solidarity, which was later crucial for the nonviolent civil disobedience that resisted government efforts to shut down the johads.
2. **Outside stimulation and facilitation:** Outsiders can be a source of fresh ideas. While action at the local level is essential, a success story typically begins when people or information from outside a community stimulate a shared awareness about a problem and introduce game-changing ideas for how to deal with it. Five young men from the group Tarun Bharat Sangh ('Young India Organization') came to the village of Gopalpura intending to set up a health clinic. But they found the greatest need was water and, on the advice of a village elder, began to work on restoring traditional earthen dams (johad) for rainwater catchment and underground water replenishment.
3. **Enduring commitment of local leadership:** Trusted and persistent leaders inspire the deep-rooted, continuing community commitment and participation necessary for success. Rajendra Singh, from TBS, maintained his role as leader and actively facilitated the construction of johads in more villages by hosting thousands of visitors to see what was achieved in Golpapura. This insisted on contracts that rigorously specifying labour and cash commitments by villages that want TBS to help them build their own johads.
4. **Co-adaption between social system and ecosystem:** Social system and ecosystem fit together, functioning as a sustainable whole. Communities create a 'social commons' to fit their 'environmental commons'. The communal-oriented

traditional Gram Sabha councils were able to manage communal enterprises such as Johad and village forests with an effectiveness not found in conventional village councils (panchayat). Young men came back home from the cities, providing additional labour for johad restoration. Villagers also organised tree planting and protection of the village forest. This mobilisation of manpower led to the restoration of the environmental support system, so that once again the ecosystem provided for people's needs.

5 **'Letting nature do the work':** Once the dams were constructed, one had only to wait for the monsoon rains. The higher water table meant that crops could grow with less irrigation, and trees could grow close enough to villages to reduce the effort for firewood collection. The recovery of forests reduced soil erosion, protecting the *johads* from siltation.

6 **Rapid results:** Quick 'payback' helps to mobilise community commitment. Results from the very first *johad* pond were seen in just a few months. During the monsoon, it filled with water and a nearby well began flowing again. This quick payback inspired more dam building. Ten years later, there were 10 such ponds in Gopalpura, holding 162 million gallons of water. The practice eventually spread to 750 other villages.

7 **A powerful symbol:** Representing the restoration process in a way that consolidates community commitment and mobilises action. The leader of TBS, the non-profit organisation stimulating these changes, became a symbol of the movement throughout the region. To underscore their commitment to the trees, villagers tied colourful *rachis* (kinship bracelets), around their trunks, a symbol of family protection.

8 **Overcoming social obstacles:** The larger socio-economic system can present numerous obstacles to success on a local scale. The restored resources – underground water, village forest and river fisheries – attracted the interest of the government, which sought to claim the resources as state property. But the 'water warriors of Rajasthan' had become well organised and were able to defend their resources.

9 **Social and ecological diversity:** With water and firewood just a short walk away, women had time to start cooperatives,

selling milk products, handicraft and soap, diversifying sources of income. Children had time to go to school, including girls who had not previously had the opportunity, bringing new skills and confidence to the village. The area of wheat fields jumped from 33 to 108 hectares, some farmers diversified into sugarcane, potatoes and onions, which increases the chance that if one crop is having a bad year, another crop is there to help the community thrive.

10 **Social and ecological memory:** Learning from the past adds to the diversity of choices, including choices that proved sustainable by withstanding the 'test of time'. Reviving the tradition of building Johads was possible because elders remembered how to construct and maintain them. The traditions of the Gram Sabha village councils, voluntary labour and foot marches ensured success and the spread of success to other villages. Nature contains an evolutionary 'memory' of its ecological design for sustainability. Because of ecological memory, the restored rivers and forests provided habitat for wildlife that had not been seen in the area for many years.

11 **Building resilience:** The forest helped to maintain and protect the watershed. Forest cover around underground water storage reduced evaporation and ensured water supply for household use and dry season irrigation even in times of low rainfall. The social organization and community solidarity were also strong protections. It was no longer necessary for women and children to haul water from distant sources. As a consequence, women had more time for housework, child care and supplemental economic activities, while children had time to return to school and the education that could provide them a more secure future.

Learning from Tarun Bharat Sangh's Dissemination of Rainwater Harvesting in Rajasthan, India; Author: Ted Swagerty

Groundwater recharge efficiency varies by location, size and shape of the rainwater harvesting structure, as well as what type of soil or rock is beneath it. Generally, bandhs are considered to provide more underground water recharge than johads. The difference between potential and actual recharge of the aquifer depends on how much water is lost as it passes downward through the soil. Water can be captured by the soil (in the process of adding to soil moisture); water can be removed from the soil by vegetation and water can flow laterally to supply streams.

1. Image: Water Warriors: Rainwater Harvesting to Replenish Underground Water (Rajasthan, India). 2. Image – tarunbharatsangh.in

References

Aditya Bahadur, E. L. (2015). Retrieved from https://studylib.net/doc/18558749/resilience-in-the-sdgs – developing-an-indicator-for-target1.5 that is fit for purpose.

Alexander, D. E. (2013). *Measuring and Building Community Disaster Resilience*. Retrieved from https://link.springer.com/chapter/10.1007/978-981-15-4320-3_9

Bakker, K., & Cameron, D. (2005). Governance, business models and restructuring water supply utilities: recent developments in Ontario, Canada. *Water Policy,* 7(5), 485–508.

Biswas, A. K. (2004). Integrated water resources management: a reassessment: a water forum contribution. *Water international,* 29(2), 248–256.

Brantz, D., & Sharma, A. (2020). *Urban Resilience in a Global Context: Actors, Narratives, and Temporalities* (p. 224). transcript Verlag.

Carlisle, S. (2013). Productive filtration: living system infrastructure in Calcutta. Landscape Urbanism.

Chakrabarty, B. (2019). Retrieved from www.orfonline.org/expert-speak/india-water-crisis-permanent-problem-which-needs-permanent-solutions-52896/.

Chandler, D. C. (2016). *The Routledge Handbook of International Resilience*. London: Routledge.

Cretney, R. (2014). Resilience for whom? Emerging critical geographies of socio-ecological resilience. *Geography Compass,* 8(9), 627–640.

Dass, M. (2011). *City with a Past – An Account of the Built Heritage of Bhopal*. Retrieved from https://architexturez.net/doc/az-cf-122795: https://architexturez.net/doc/az-cf-122795

Eraydın, A. T.-K. (2013). *Resilience Thinking in Urban Planning*. Dordretch, London: Springer.

https://economictimes.indiatimes.com. (2016, March 20). Retrieved from https://economictimes.indiatimes.com/news/politics-and-nation/vanishing-wetlands-indiscriminate-development-poor-regulation-are-wrecking-a-critical-piece-of-indias-ecology/articleshow/51473398.cms

https://niti.gov.in/content/composite-water-management-index. (2018, June). Retrieved from www.niti.gov.in: https://niti.gov.in/content/composite-water-management-index-june-2018–0

Mrunmayi Wadwekar, A. W. (2018). *Urbanisation and Environment: A Case of Bhopal*. Retrieved from www.researchgate.net.

Rijke, J., Farrelly, M., Brown, R., & Zevenbergen, C. (2013). Configuring transformative governance to enhance resilient urban water systems. *Environmental Science & Policy,* 25, 62–72.

Schlüter, M., & Pahl-Wostl, C. (2007). Mechanisms of resilience in common-pool resource management systems: an agent-based model of water use in a river basin. *Ecology and Society,* 12(2).

Sudmeier-Rieux, K. I. (2014). Resilience–an emerging paradigm of danger or of hope? *Disaster Prevention and Management*.

TeamDB, P. (2016). *The City of Vanishing Lakes, Bhopal*. Team DB Post.

White, I. (2010). *Water and City: Risk, Resilience and Planning for a Sustainable Future*. Retrieved from Routledge, London, UK. https://doi.org/10.4324/9780203848319

www.hindustantimes.com/analysis/india-must-not-look-at-its-water-crisis-in-isolation. (2019, November 13). Retrieved from www.hindustantimes.com/analysis/india-must-not-look-at-its-water-crisis-in-isolation

www.rainwaterharvesting.org/bhoj_lake/bhoj_lake.htm. (2009). Retrieved from www.rainwaterharvesting.org/bhoj_lake/bhoj_lake.htm

5

SOUTHERN SOCIOECOLOGICAL RESILIENCE

Theorising a New 'Normal'

Sandeep Balagangadharan Menon

Beyond Ecological Resilience | Socioecological Resilience: Global Shifts in Understanding

Discourses on urban resilience in the Global South in the present epoch of Anthropocene cannot be limited to human-centric constructs of sociopolitical and economic concerns. They need to have a more nuanced understanding of the questions concerning the local bio-geo-climatic situation as well as the effects of the global climate crisis. In the last ten thousand years of Holocene, we humans have evolved from the life of a hunter-gatherer to being a part of the present urban age. However, it is important to note that this shift was possible largely due to the steady temperature range of the world in the Holocene. There have been no glacial periods or extreme warming of the world in this period. This allowed for a certain sense of probability in the climatic events, seasons became steady and cyclical with a fair degree of predictability. An efficient agricultural system was devised by humanity, refining it over time to feed the growing human population. The dawn of the civilisations, as well as urbanity, owes its genesis to the predictable climatic range of the Holocene. Human brains are cognitively programmed to be social. This ability of the species to congregate and work in conjunction with the others has paved the way for all the advancements humanity has achieved. However, specific lifestyle choices, habitable space configurations, materiality, cultural practices and dietary choices have always been shaped as responses to the biological and climatic signatures of the geographic regions where the communities live.

The large-scale interventions triggered by humanity in the last two hundred years, post the industrial revolution and especially in the era of the great acceleration, post-1950 (see box) have started affecting atmospheric

DOI: 10.4324/9781003098461-5

warming. This phenomenon of global warming is now threatening the predictability of climatic events. The current pandemic situation of the novel coronavirus has been attributed to human actions of encroachments into wildlife habitats, thereby increasing risks of zoonotic transitions (Keusch et al., 2009).

There is scientific evidence on the anthropogenic greenhouse gas emissions causing climate change and ocean acidification, increasingly threatening the resilience of the biomes and the human societies that depend on them (Kresge Foundation, 2017; Malhi et al., 2020).

This chapter is divided into eight sections. The second section discusses the socioecological peculiarities of Indian cities. These cities which have been lived in for centuries and have transformed over time to accommodate new layers of urbanisation are excellent case studies for discussing the concept of socioecological resilience. The third section discusses the proposed method of inquiry adopted for the study. The main research question of the study is aimed at understanding the patterns of discord in the way Global South coastal cities have developed with scant regard to the natural processes. The fourth section introduces the city of Kochi as a case example for the study. The fifth section looks at the paradox of urbanisation and the environment. The discussion is carried forward through the historic evolution of the city and how anthropocentric activities alienated the urban form from the bio-geography of the estuarine edge. The sixth section brings to light the vanishing cultural practices which were prevalent in the region till the turn of the century. These practices had a significant role in maintaining the balance of human actions and their implications on the natural processes and vice versa. Section 7 continues this discussion to document the present scenarios of the development of the city. The concluding section sums up the study and the way forward, suggesting effective management strategies based on the learnings from the socioecological systems.

Box 5.1 The Great Acceleration

The term Great Acceleration (GA) refers to the quick surge in the magnitude of the varied human activities approximately in the decade of 1950s, and therefore, there is a subsequent increase in the disruption of diverse ecological processes and patterns of the earth's natural systems. The International Geosphere-Biosphere Programme (IGBP) initially published the 'Great Acceleration indicators', in the IGBP synthesis, Global Change and the Earth System in 2004. Since then they needed an update and reassessed the indicators with the technical assistance of the Stockholm Resilience Centre (SRC). The research highlights the 'Great Acceleration' in human activity from the start of the Industrial Revolution in 1750–2010 and the increasing changes in the Earth System. The research outcomes and data published in the official website www.igbp.net point to the fact that human activities, especially global economic-related activities, have affected the massive deteriorations in the earth systems. The study shows key evidence that the earth's geo-climatic systems are moving beyond the variability range within which they operated in the last 10,000-year-old epoch of the Holocene. It was this geo-climatic certainty of the Holocene which enabled humankind to pursue agriculture and lead to a stable life (News: IGBP Global Change Website, 2015).

Source: Author

Adapting cities to the threats of climate crisis often demands granular changes in the systems as well as offers opportunities to the urban systems. The prominent issues on this in the Global South are increasing densities of population and rapid often unplanned growth of cities. One of the predominant challenges of adapting to climate change revolves around questions of uncertainty and limitations in knowledge since future predictions can often be complex and marked by uncertain cause–effect relationships and may have inconsistent scientific evidence (Deppisch et al., 2011; Harte, 2009).

The fact that humans are a part of the natural system is not new knowledge, but the need for emphasising this is even more necessary in the present times. This integrated approach would mean that the social and ecological systems are coupled and are interdependent. The term 'socio-ecological system' was coined by Firkit Berkes and Carl Folke in 1998 as a means to give equal importance to the two aspects of social and ecological fields of study for research purposes. Berkes and Folke in their seminal book 'Linking Social and Ecological Systems: Management Practices and Social Mechanisms for Building Resilience' argue that the conventional scientific knowledge and political policies aimed at seeing nature as a mere '*ready-for-extraction*' resource disregard the understanding that ecosystems could be nonlinear and multistable. They identify systems theory as a method to unravel management puzzles and the adoption of adaptive management to link science and ecological management policy (Berkes & Folke, 1998). Hence, discussions on resilience concerning socioecological systems are centred on the ideas of *adaptation, learning, self-organization, and the general ability to persist disturbance* (Folke, Resilience: The emergence of a perspective for social-ecological systems analyses, 2006)

With the advent of globalisation in India, the tussle between developmental programmes which address human well-being and those which address environmental services is often at loggerheads. This dilemma of achieving improved human conditions at the cost of degradation of ecosystems has been a trend historically (Turner, 2008), but it is more pronounced in the mid-sized historic cities in India.

Also building resilience against all odds may not be possible in the tropical cities which are at the intersection of the wrath of the climate crisis and the unprecedented pressures of development and rapid urbanisation. Some solutions that may deem appropriate to ensure resilience against certain climatic vagaries may end up pushing some parts of the city and communities residing there to be more vulnerable. This could entrench and exacerbate the inequalities in the city (The Seven Principles-Introduction, n.d.).

Like all living systems, landscapes also comprise three basic characteristics: 'landscape structure', comprising the spatial arrangements of the various entities within it; 'landscape function' which includes movements and flows of materials, organisms, energy in the system; and 'landscape change' which refers to alterations in the landscape structure or function over time (Dramstad et al., 1996).

Section 2: Southern Cities – Socioecological Peculiarities of Indian Cities

Indian cities vary vastly from those in the Global North. This is due to the bio-geographic peculiarities of the urbanised zones in the Global South clubbed with the socio-economic opportunities they provide to the larger hinterlands, thereby triggering a steady influx of intra-national migration of people. Richard Plunz elucidates in his book City Riffs, 2017 that urbanism should be viewed as the outcome of an integration of design, ecology, and engineering integrated with influences on urban cognition such as social, economic and psychological interactions (Plunz, 2017). Recent thrust on cities as 'ordinary cities' (Robinson, 2006) allows for the reading of the otherwise invisible mid-sized cities of the Global South as urban centres which deserve a place in the research spectrum. Most Indian cities have been inhabited for centuries, their historic urban cores often shaped as a response to the natural processes, the lay of the land[1] and the land–water interactions. But mid-sized historic cities in India are on the cusp of rapid unregulated urbanisation aided by a gamut of government-led infrastructural projects which are often guided by misplaced political aspirations. These often ignore the grounded realities of the ecological signature of the cities wherein they are implemented. Such developments render the cities highly vulnerable to climatic and geological catastrophes. The coastal cities of the Indian peninsula are emblematic of this. These cities have seen an increase in both the frequency of incidence and the intensity of climatic vagaries like coastal flooding, hurricanes and increased erosion due to rising sea levels and increased wave action. Interestingly, the past decades have also seen their population increasing steeply as compared to the earlier century.

The focus of resilience thinking by the governmental bodies in the Global South is also majorly steered towards short-term recovery post-catastrophe, focusing on immediate results. This myopic view of addressing resilience by disregarding the long-term stress may lead to short-term robustness of the system, but it may not be amenable to allow for either adaptability[2] or transformability[3] (Smith & Stirling, 2010).

Section 3: Proposed Method of Inquiry-Inductive Research and Systems Thinking

The patterns, resemblances from the observations/premises from the site, inform the formulation of the conclusion or generation of the theory. The main research question of the study is aimed at understanding the patterns of discord in the way Global South coastal cities have developed with scant regard to the natural processes and to the ecological signatures of the geography where they are situated. The chapter intends to formulate a pragmatic approach to incorporate socioecological resilience thinking for coastal cities. The proposed method of inquiry adopted is based on the theory of 'systems thinking' since it attempts to comprehensively addresses issues as complex as those that one finds in rapidly urbanising geographies. An ecological approach to such a framework should also be based on the landscape strategy as one that is interlinked to the living beings (species), their habitats (including the abiotic entities) and the cultural interdependencies of the human populations inhabiting the region (Cantrell & Holzman, 2016). A conceptual framework for assessing socioecological resilience in the coastal cities of the Global South could prove to be beneficial in understanding the complexities of vulnerabilities faced by these cities.

Traditionally, frameworks for ecosystem studies have emphasised the correlation between human activities and biogeophysical drivers, shaping up the ecosystem dynamics. Even though this model acknowledges the social and the environmental aspects, it is ineffective in getting a complete understanding as it excludes interactions and feedbacks between the domains. It is these interactions and feedbacks that often lead to long-term changes in the ecosystems (Redman et al., 2004).

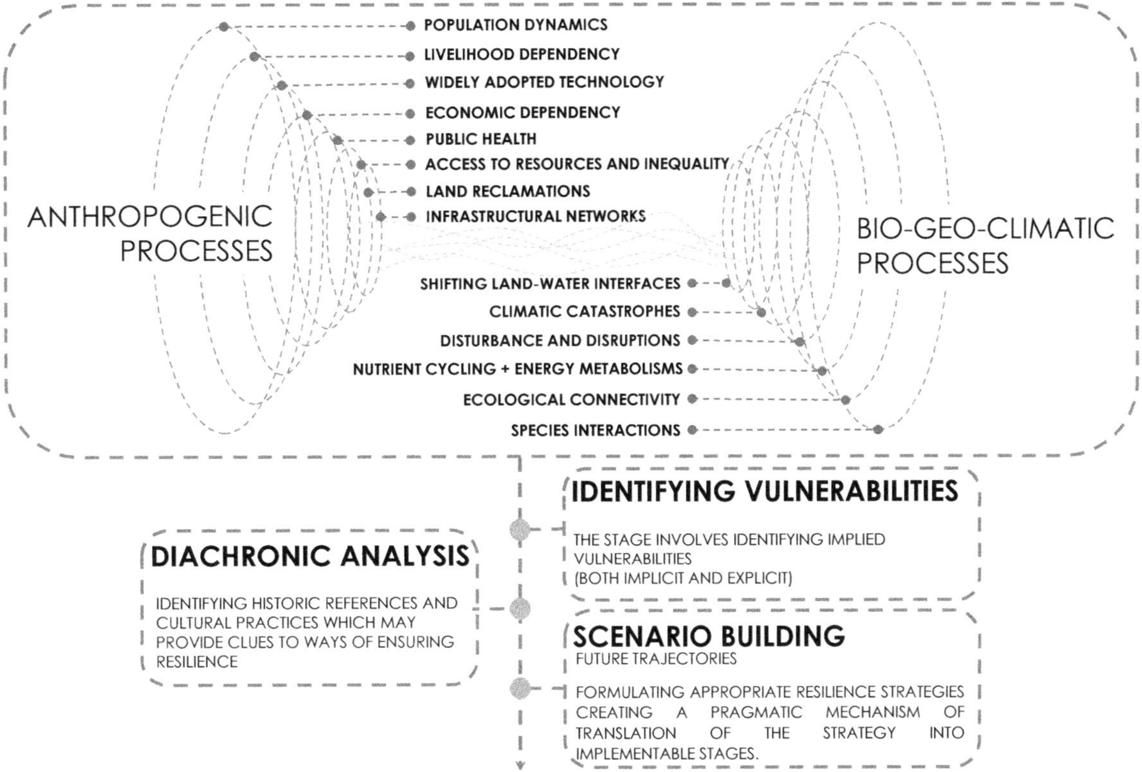

Figure 5.1 The framework examines overlaps, interdependencies and interactions, stakeholders and structural manifestations between these aspects and tries to examine them across time, scale and systemic capabilities.

Source: Author

The proposed conceptual framework involves three distinct stages. The first stage involves identifying the various dependencies listed under the broad categories of human-induced processes and bio-geo-climatic processes.

The dependencies listed under the human-induced processes are population dynamics, livelihood dependency, governmental policies, widely adopted technology, economic dependency, public health and nutrition-related aspects, access to resources and inequality, land reclamations and infrastructural networks. The dependencies listed under the bio-geo-climatic processes include primary production, species interactions, connectivity, nutrient cycling and energy metabolisms, disturbance and disruptions, precipitation and heat-related climatic catastrophes, and land–water interfaces.

The framework examines overlaps, interdependencies and interactions, stakeholders, structural manifestations between these aspects and tries to examine them across time, scale and systemic capabilities.

The second stage involves analysing the system using the following steps: diachronic analysis of the socioecological systems to identify historic references and cultural practices which may provide clues to ways of ensuring resilience. The stage also involves identifying implied vulnerabilities (both implicit and explicit).

The final stage of the process involves scenario building to find future trajectories which will help in formulating an appropriate resilience strategy and help in creating a pragmatic mechanism of translation of the strategy into implementable stages.

Section 4: Shifting Dynamics of Systemic Ecologies – Case Study: Kochi

The city of Kochi in Kerala offers a kaleidoscope of ecological issues and potential opportunities representative of a coastal mid-sized historic city in the Global South. The largest city in the southern state of Kerala, Kochi is the centre of commercial and economic activities in the state.

The city owing to its prominent location in the Arabian Sea just a few miles north of the Indian Ocean is also a historically important port. This topography of the city is characterised by comparatively flat terrain and includes a large expanse of estuarine backwaters of the Vembanad-Kol wetland estuarine system, multiple rivers and canals with the urban areas spread across several islands scattered in the backwaters. The city was traditionally an amphibious entity dwelling on both land and water. The sinuous waterways which separated the landmasses were ecologically productive landscapes and they played the role of connecting the various communities that dwelled on various islands. These waterways were exploited by the early settlers as established routes of movement between the islands. The land–water edge in these monsoon-drenched landscapes was never static lines as one would read out of a map but morphed and shifted depending on the currents, tides and rainfall received in the verdant west-facing slopes of the Western Ghats mountains. Vembanad-Kol wetland system fed by ten rivers, covering an area of 1,512 sq. km, is the largest estuarine system in the Indian peninsula. It is a biodiverse landscape and has sub-fossil deposits (Ramsar, 2002); this estuary has two seasonal openings into the Arabian Sea – one at Parur (north) and the second at Andhakaranazhi – and a permanent opening at the mouth of Kochi.

During floods, seawater tides enter the estuary via Cochin bar mouth (12 m depth) and the flow reverses during the ebb tide. Seven major rivers (Chalakudi, Periyar, Muvattupuzha, Meenachil, Manimala, Pamba and Achencoil) discharge freshwater into the estuary perennially.

Backwaters as ecosystems are one of the largest commons of the state, and they play a dominant role in the economy of Kerala. Despite this, its economic importance has not been properly recognised either in the academic or policy realms (Thompson, 2001).

Box 5.2 Story of Vembanad-Kol wetlands

The Vembanad-Kol wetlands once covered an area of 2,033.02 sq. m and its river basins spread over 6,126.48 sq. km area. The area of the Vembanad Lake during 1917, 1970 and 1990 had declined to an extend of 290.85 sq. km, 227.23 sq. km and 213.28 sq. km, respectively (Gopakumar & Takara, 2009).

A total of 63.62 sq. km area had reclaimed from the lake during the period 1917–1970 and was primarily for the formation of polders and to enlarge the extent of the Wellington Island and Cochin Port.

Source: Harikumar P.S. (2016) http://wgbis.ces.iisc.ernet.in/energy/lake2016/Plenary/T5_DrHarikumar_abs.pdf accessed on 01.05.2020

Historically, the region owes its form to a series of transgression and recession of the sea and the land. The Kerala coast originally had three major physiographic divisions. They are

- High lands
- Midlands and
- Low lands

The Vembanad estuary lies in the western lowlands. The land originally is believed to have been thickly forested which was buried under the sea due to a sudden catastrophe like an earthquake. Traces of this buried forest have been excavated from the Kari lands of Kuttanad (south of Kochi city).

The trading port before the 14th century was the Port of Muzhiris (present-day Kodungallor) on the north of the city of Kochi. It is believed to have had trade connections with ancient Rome and Greece. In the year AD 1341, a heavy flood silted up the Muzhiris harbour and pushed open a harbour mouth to the south at the present location of the Kochi Harbour (Logan, 2000). This pushed the development of a new town at the island of Mattancherry on the western sand bar of the estuary. The settlement comprised a diverse mix of people from various countries. The new town centre of Kochi started developing on the eastern banks of the estuarine zone in the mid-17th century.

An extensive survey and settlement of the wetlands called the '*Kandezhuthu*' was conducted in the year AD 1821 (996 Malayalam Era), and the survey and settlement records of garden land were done during the period AD 1837–1838. Extensive reclamation of the backwater started between 1879 and 1889 when the Ernakulam foreshore was greatly improved in terms of public spaces, public buildings and infrastructural facilities.

The introduction of Railways in 1905 was also a milestone in the evolution of the city. Kochi was gradually developing into an administrative city as well during the 19th and 20th centuries. The development of the Kochi Port by dredging the channels to allow for deep-sea vessels to enter the harbour in all seasons in the 1920s spearheaded by Robert Bristo led to the formation of the man-made island of 780 acres called the 'Wellington Island'. The commissioning of Cochin Port as a year-round port was responsible for the fast development of Ernakulam and suburban areas.

The post-independent developments in Kochi saw it emerge as a major commercial port, the Greater Kochi Region, encompassing 731 sq. km. The coastal areas are densely populated with a density of 6,300 persons per sq. km in the city compared to the average density of 819 persons per sq. km in the state (Corporation of Kochi, Govt. of Kerala).

The city has seen an unprecedented increase in its population and the rampant construction of buildings and allied infrastructure in the last two decades. This is primarily symbolic of the trends in most tier two cities of India post the globalisation and the increase in the upwardly mobile middle class in India. A cursory survey of the city reveals that the development does not take into cognizance the ecological peculiarities of the region. This makes the city highly vulnerable to climate change-induced sea-level rise and catastrophes like storm surges. The city despite its 'invasive' urbanisation still has patches of relict ecologies that are intact and has communities

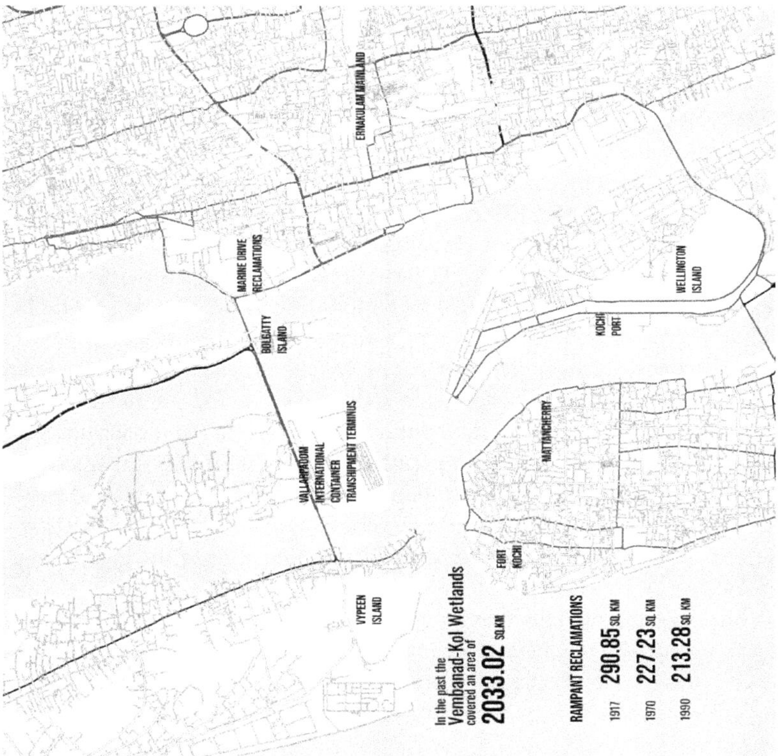

Figure 5.2 Kochi city straddles the Vembanad estuarine zone and its many islands.

Source: Author

whose livelihoods depend on the seasonal variations. The presence of this unique contradiction in the way humans shape the environment and the environment shapes the way humans live is prominent in most southern coastal cities. The haphazard nature of the development and the scant regard for natural processes only make the situation worse.

Section 5: The Paradox of 'Urbanisation' and 'Environment' – Identifying the Vulnerabilities

The Government of Kerala approved a Development Plan (DP) for the central city of Kochi in the year 1961, and this has undergone subsequent revisions. The zoning regulations of the structure plan were revised in the year 2007. But these were not seen as satisfactory to help shape the trajectory of urbanisation in the city and its peripheries. Hence a revised Development Plan for Kochi city region was prepared in 2010 for the year 2031 (Department of Town and Country Planning, 2010). Under the aegis of the 'Centre for Heritage, Environment and Development under the Kochi Municipal Corporation, Kochi's first master plan was prepared with a vision period of 20 years. The master plan was seen as a multidimensional exercise which looked at the regional framework, physical and social infrastructure, environment, heritage, tourism and resource mobilisation, etc. The master plan was prepared to guide the constituent local governments within the city region to formulate their individual development policies in a coordinated and informed manner. This was imperative to note that the recent spate of development in Kochi has not taken into consideration the interplay of natural processes since environmental planning was considered non-existent by the governmental bodies responsible for the planning of the city (Benjamin, 1998).

The bio-geo-climatic characteristics of the city need to be understood to assess its potential to withstand various vulnerabilities. The physiography of the city and its suburban agglomeration can be divided into two distinct geological zones running in the north–south axis (Menon, 1965):

1 The relatively flatter coastal western plains interspersed by numerous wetlands
2 The eastern lateritic low hills

The eastern low hills are a part of the Midlands of Kerala characterised by laterite capped on an archaean[4] rocky substrata like gneiss or charnockite running in the north–south direction parallel to the Western Ghats mountain ranges and the coastal edge. This zone has multiple drainage basins and rivers and streams crisscrossing the topography originating from the east and flowing towards the flatter west. Owing to the presence of lateritic soil and parent rock substrata, these zones are suitable for urban development. Laterite is also a traditional building material in Kerala and is quarried out of these low hills. However, the rampant quarrying of laterite and the hard-crystalline rock from these areas have seen severe environmental imbalances in the recent past (Maya et al., 2012).

The western plains consist of recent depositions of alluvium interspersed by tidal canals and wetlands. The elevation of this zone is in the range of a metre above the mean sea level (Benjamin, 1998). This makes the zone highly prone to sea-level rise. This is also the most densely urbanised part of the city with high rises and rampant reclamations of the backwaters. The western edge of the city is marked by densely populated islands and sandbars. The geological strata of sediments are in layers of sand, clay, clayey sand, or sandy clay with a band of laterite of varying thickness. Soil exploration detected that this deposit is present evenly to a depth of about 50 m from the sea level. This makes this zone highly prone to sinking and geological instability in the case of a seismic event that may be aggravated by human activities (Department of Town and Country Planning, 2010).

There are records of geological instability reported during the construction of the Cochin Port when massive quantities of soil were dredged and moved around to deepen the harbour and create a man-made island of Willingdon Island. An earthquake was reported on 15 January 1934 at 2.15 pm which resulted in the subsidence of the buildings of the Willingdon Island (Benjamin, 1998). The epicentre of this quake was surprisingly 2,900 km away in the Himalayan valleys in Nepal. This earthquake dubbed the *Magh* Earthquake[5] or the 'Great Nepal-Bihar Earthquake' is one of the worst in the history of the subcontinent (DownToEarth, 2015). Incidentally, there have been reports of quake tremors felt in the high rises of Kochi on 25 April 2015 as well when another massive earthquake shook the Kathmandu Valley again (DC Correspondent, 2015). The increased reclamation of the backwaters and rampant construction of high rises in the western plains and sand bars make the urban development vulnerable to seismic stress.

The land–water edges of Vembanad have always been fluctuating and dynamic based on seasonal variation. The occupation and cultural practices of the inhabitants have been traditionally linked to this amorphous water edge. Islands and sand bars were inhabited by the fisherfolks and the river flood plains with freshwater supply were inhabited by the rice paddy cultivators. The rest of the land was brought under plantations of coconut, areca nut, etc. Till about seven decades ago, agriculture was the primary source of income for the state. All this changed once India gained independence, and the thrust was laid on development. Kochi was fast emerging as a financial hub of Kerala owing to the presence of the port.

Rampant reclamations have happened in the past century. Many government schemes encouraged reclamations, including the extension of the city downtown Central Business District called the 'Marine Drive' with a concrete pavement boulevard marking the edge of the waters was reclaimed from the estuary by dredging the backwater floor and depositing the soil at the land's edge. About 50 acres of land was 'reclaimed' in the 1970s (Anand, 2013), and the Greater Cochin Development Authority (the govt body responsible for the city development) touts it as an achievement in urban development in the history of the city. In August 2018, the city of Kochi was battered by an unprecedented amount of rainfall owing to severe floods in the city and many parts of the state. The extreme flooding in the city was primarily attributed to the reduction in the water carrying capacity of the estuary. The Planning Commission report of July 2008 had documented the tremendous reduction of the carrying capacity of the Vembanad Estuary – currently about 0.6 billion cubic metres from the earlier 2.4 billion cubic metres primarily owing to rampant land reclamation (Kuttoor, 2018). The total annual sediment yields the estuary bears each year is about 32 million tonnes/year. The mean depth of the lake has reduced from 6.7 m to 4.4 m (Padmakumar et al., 2002). Despite knowing this, the city continues to grow into the wetlands.

The land is seen as a 'resource' and an 'investment' in Kerala. It has been a cultural practice even in precolonial times as well. However, the earlier generations valued the forces of nature and most of the rituals of handling land were in tune with nature. The post-liberalisation decades have seen a tremendous increase in the number of people migrating to Kochi for want of better living standards and occupational opportunities. This puts immense pressure on the land, and the prices are driven up which makes it an even more lucrative investment opportunity for the locals. This notion needs to change since it indirectly makes the place more vulnerable. All other aspects of land are pushed away, and the land is considered as a place to build only.

There are also reports of the reduction of biodiversity in the Vembanad estuary and the surrounding lands due to widespread pollution and urbanisation. The Vembanad Life Region supports about 1.6 million people who depend on it for their livelihoods. It also supports India's third-largest wintering waterfowl population. It provides habitats for many endangered fish species and mangroves (Ghosh, 2015). Other than the issues brought about by urbanisation, modern unscientific fishing practices are also wrecking the ecological balance of the areas. Overfishing, use of nets with low porosity, building barriers that inhibit migration of species, overexploitation of houseboats and pollution loads are all adversely affecting the biodiversity of the region (Remani et al., 2010).

Box 5.3 Urban policy and governmental role in the protection of the estuarine edge

In The past water bodies, as large as the wetlands of Vembanad, were always considered as common resources and as public spaces. Alluding to protection to them while extracting resources from them was considered a collective responsibility of the communities dependent on the waterbody. As colonial rule took over parts of India, so did the European visions of legislative reforms. It was in the Indian Forest Act 1927 that the first recorded evidence of planning interventions for ecological assets was formalised.

The Water (Prevention and Control of Pollution) Act of 1974 called for the prevention of the flow of untreated sewage and industrial effluents into water bodies. In 1982, India became a signatory to the Ramsar Convention on Wetlands, an international convention that focuses on the conservation of wetlands. Vembanad was inducted as a Ramsar Wetland site in 2002. A National Wetland Conservation Programme (NWCP) was formed by the Union Government in 1985 under the aegis of the Ministry of Environment and Forests under which 115 wetlands were identified and conservation and management initiatives were issued (Ramsar, 2002).

In 2016, the National Lake Conservation Plan and the National Wetlands Conservation Programme were merged to form the National Plan for Conservation of Aquatic Ecosystems (NPCA). The programme intends to ensure the holistic conservation of wetlands through an integrated and multidisciplinary approach (Jainer, 2020).

Section 6: Vanishing Cultural Practices

Cities like Kochi are unique in many ways that there have been prevalent cultural practices within the region that were attuned to operating in the bio-region. Many of these practices have been instrumental in shaping the eco-social systems of the city. An understanding of these may serve us with cues to ascertain the way forward.

The Pokkali System of Cultivation is one such socioecological practice that was prevalent in the estuarine low-lying lands of central Kerala. The Pokkali strands of rice are saline resistant, and this allows for the communities living along the estuary to cultivate this rice during the monsoon months (April–November) when the salinity gradient of the water is lower

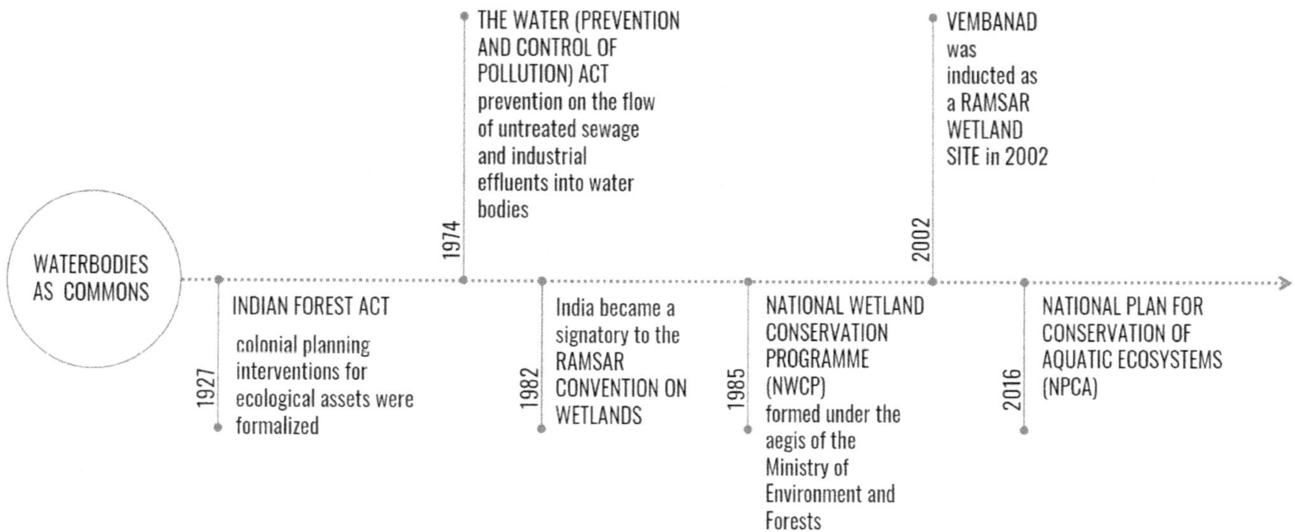

Figure 5.3 Urban policy and governmental role in the protection of the estuarine edge.

Source: Compiled by Author

than the rest of the months. During the rest of the year, prawn cultivation is encouraged in the same space. The high tide brings in the young prawn, and this is caught in the bunds. Often a kerosine oil lamp is hung on the sluice gate to 'attract' prawn fry (Kutty, 1987). During the neap tides, the prawns in the fields by allowing to let the water to escape out through the netted-sluice gate. The prawns are harvested from December till summer every fortnightly usually following the lunar phases. This age-old practice works on the principles of modern-day permaculture. The parts of the plant, left behind after harvesting the grain, decay and become nutritious food for the prawns. At the end of the summer months when the Prawns are harvested, the carapace and the rest of the animal residues form manure for the Pokkali rice strands.

Sacred groves are fragments of heavily wooded landscapes protected by cultural beliefs, taboos and customs of the local community. They are an expression of the relationship of humans with the 'divine' or with nature (Hughes, 1998). In Kerala, they are referred to as 'Kaavu'. They usually consist of multispecies, multitiered patches of primary forests maintained in a relatively undisturbed condition for centuries. They highlight the climax stage of vegetation which tends to be richer in species diversity than the other stages of its ecological succession. Most sacred groves harbour and protect numerous endemic species which are on the verge of being endangered or being extinct. Despite being heavily fragmented and surrounded by the newly urbanised areas, they play an important role in continuing the ecological integrity of the region. Over the past few decades, Kerala has witnessed a drastic reduction in the number of its groves due to changing sociocultural beliefs.

These are some examples of sociocultural practices which work as a response to the environmental processes in the region. Many of these practices are slowly dying out and are replaced by technological advancements.

Section 7: The Present Condition

The state of Kerala is marked by a high literacy rate (Ministry of Statistics and Programme Implementation, 2020), and the Human Development Index in the state of Kerala is considered a model for the Global South. The state also has a high inflow of income from the expatriates of the state who work in the Middle Eastern countries. However, these high indices do not necessarily mean an ecologically sensitive development. Kerala is one of the first states in India to allow for power devolution to local bodies and also allow for participatory planning. This is often seen as an imitable model. However, the most glaring gaps in this development trajectory are probably due to the lack of political will to consider the peculiarities of the natural environment in the state. Most master plans are not implemented properly, and most urban local bodies do not have control over the proposed plans. This is plagued by the lack of coordination between the multiple organisations. Well-intentioned urban projects which herald concepts like blue-green urbanism, thrust on cycling/pedestrianisation, etc., are too sporadic and many a time ends up being 'tokens'. Kochi has been selected as one of the 'Smart Cities' under the Government of India's Smart Cities Mission (SCM). The city could benefit through the SCM under its Area Based Development (ABD) strategies. The geographic vulnerability of the city is on the estuarine edge which is susceptible to flooding needs to be taken into consideration while planning for the future for the city.

Section 8: Conclusion – Suggesting Effective Management Strategies Based on the Learnings from the Socioecological Systems

In this chapter, we discussed that resilience discourse for cities in the Global South needs to be more nuanced and requires a more grounded approach than those of the cities of the Global North. A three-stage methodological framework was put together to discuss how a coastal city could be analysed. The case of a medium-sized historic coastal city of Kochi was used to discuss ways of reimaging the future by visiting some of these crucial interlinkages between ecology, culture and communities. Certain policy measures are underway, but they do not pay much attention to many of the vanishing cultural practices like Pokkali farming and sacred groves which have contributed to sustaining natural ecologies for centuries. The possible way forward could be addressed by mediating urbanisation with local ecologies, resulting in reduced vulnerabilities and better livelihood dependencies on the natural systems.

Addressing the city as a rhizomatous hyper-connected network of biotic and abiotic interactions opens up the possibilities of redefining the 'southern socioecological resilience' in a more granular manner. Effective management strategies could then be built upon from the learnings from the socioecological systems that could allow for integration of biological diversity, ecological and economic security, and also help shape a future trajectory for the city and its peri-urban ecologies.

Notes

1 The 'lay of the land' refers to the topographical characteristics of the land, indicating whether the land is sloping gently or whether the slope is steep. The soil conditions, steepness of the slope of the land and the impending rainfall of the region are important factors of soil erosion.

2 Adaptability is defined as '*the ability of a system to reorganize and reconfigure to cope with disturbances without losing its functional capacity and systems identity*'. Since cities are human-dominated ecosystems, adaptability depends on individual human actions as well as community practices of management of the natural systems like agriculture and reclamation (Walker et al., 2004).
3 Transformability is defined as *'the capacity of a system to create a fundamentally new system when ecological, economic, or social (including political) conditions make the existing system untenable'* (Walker et al., 2004).
4 Archaean rocks, also known as Pre-Cambrian rocks, are one of the oldest rocks in the geological history of the earth. Most archaean rocks are azoic (devoid of any forms of remnants of life).
5 Magh refers to the name for the season comprising the months of January–February.

References

Anand, S. N. (2013, May 30). *The City's Posh Promenade*. Retrieved from The Hindu: www.thehindu.com/features/friday-review/history-and-culture/the-citys-posh-promenade/article4766107.ece

Benjamin, P. (1998). *Environmental Resource Assessment of Cochin*. Kochi: School of Environmental Studies, CUSAT.

Berkes, F., & Folke, C. (1998). *Linking Social and Ecological Systems: Management Practices and Social Mechanisms for Building Resilience*. New York: Cambridge University Press.

Cantrell, B., & Holzman, J. (2016). *Responsive Landscapes: Strategies for Responsive Technologies in Landscape Architecture*. Oxon: Routledge.

Corporation of Kochi, Govt of Kerala. (n.d.). *Kochi Development Plan*. Kochi: Govt of Kerala. Retrieved from https://cochinmunicipalcorporation.kerala.gov.in/documents/10157/73076ec1-6197-435a-8973-4dd5626e0225

DC Correspondent. (2015, April 26). *Nation | Current Affairs: Deccan Chronicle*. Retrieved from Deccan Chronicle: www.deccanchronicle.com/150426/nation-current-affairs/article/earthquake-quake-vibrations-rock-kochi

Department of Town and Country Planning. (2010). *Development Plan for Kochi City Region 2031*. Kochi: Government of Kerala.

Deppisch, S., Hasibović, S., & Albers, M. (2011). Plan B:altic: A social – ecological approach to climate change adaptation. In K. Otto-Zimmermann (Ed.), *Resilient Cities: Cities and Adaptation to Climate Change Proceedings of the Global Forum 2010* (p. 157). Bonn: Springer Science+Business Media.

DownToEarth. (2015, July 4). *News: Major Past Earthquakes in Nepal*. Retrieved from DownToEarth: www.downtoearth.org.in/news/major-past-earthquakes-in-nepal – 49546

Dramstad, W. E., Olson, J. D., & Forman, R. T. (1996). *Landscape Ecology Principles in Landscape Architecture and Land-Use Planning*. Washington, DC: Havard University graduate School of Design, Island Press and the American Society of Landscape Architects.

Folke, C. (2006). Resilience: The emergence of a perspective for social – ecological systems analyses. *Global Environmental Change*, 253–267.

Folke, C., Biggs, R., Norstrom, A. V., Reyers, B., & Rockstrom, J. (2016). Social-ecological resilience and biosphere-based sustainability science. *Ecology and Society*, 21(3).

Ghosh, P. (2015, September 11). *Vembanad Lake is Showing Us the Future of Conservation in India*. Retrieved from Scroll.in: https://scroll.in/article/751302/vembanad-lake-is-showing-us-the-future-of-conservation-in-india

Harikumar P. S. (2016). http://wgbis.ces.iisc.ernet.in/energy/lake2016/Plenary/T5_DrHarikumar_abs.pdf accessed on 01.05.2020.

Harte, J. (2009). Numbers matter: Human population as a dynamic factor in environmental degradation. In L. Mazur (Ed.), *A Pivotal Moment: Population, Justice and the Environmental Challenge*. Washington, DC: Island Press.

Holz, E. a. (2017). *Trends in Urban Resilience*. UN Habitat, Nairobi.

Hughes, J. (1998). Sacred groves around the earth: An overview. In P. S. Ramakrishnan (Ed.), *Conserving the Sacred for Biodiversity Management*. New Delhi: Oxford & IBH Publishers Co.Pvt.Ltd.

Jainer, S. (2020, January 3). *Down to Earth -Urbanisation Section*. Retrieved October 10, 2020, from www.downtoearth.org.in/blog/urbanisation/how-do-india-s-policies-and-guidelines-look-at-urban-lakes – 68662

Kerner, D. A., & Thomas, S. J. (2014). Resilience attributes of social-ecological systems: Framing metrics for management. *Resources*, 672–702.

Keusch, G. T., Pappaioanou, M., González, M. C., Scott, K. A., & Tsai, P. (Eds.). (2009). *Sustaining Global Surveillance and Response to Emerging Zoonotic Diseases*. Washington, DC: National Academies Press. Retrieved from www.ncbi.nlm.nih.gov/books/NBK215318/

Kresge Foundation. (2017). *Bounce Forward: Urban Resilience in the Era of Climate Change*. Washington, DC: Island Press.

Kuttoor, R. (2018, September 24). *Low Capacity of Vembanad Contributed to Flood: CWC*. Retrieved from The Hindu: www.thehindu.com/news/national/kerala/low-capacity-of-vembanad-contributed-to-flood-cwc/article25031626.ece

Kutty, M. (1987, April). *Fish Culture in Rice Fields*. Retrieved from Food and Agriculture Organisation of the United Nations: www.fao.org/3/AC180E/AC180E10.htm

Logan, W. (2000). Malabar Manual (Vol. 2). *Asian Educational Services,* New Delhi; Chennai.

Malhi, Y., Franklin, J., Seddon, N., Solan, M., Turner, M. G., Field, C. B., . . . Nancy. (2020, January 27). Climate change and ecosystems: Threats, opportunities and solutions. *Philosophical Transactions Royal Society B*, 375(1794). https://doi.org/10.1098/rstb.2019.0104

Maya, K., Santhosh, V., Padmalal, D., & Aneesh Kumar, S. R. (2012, July). Impact of mining and quarrying in Muvattupuzha River basin, Kerala- An overview of its environmental effects. *Bonfring International Journal of Industrial Engineering and Management Science*, 2(1). Retrieved October 25, 2020.

Menon, S. A. (1965). *Gazetteer of India-Kerala-Ernakulam*. Trivandrum: Kerala Gazetteers | Government of Kerala.

Ministry of Statistics and Programme Implementation. (2020). *Household Social Consumption on Education in India*. New Delhi: National Statistical Office.

News: IGBP Global Change Website. (2015, January 15). Retrieved from International Geosphere Biosphere Programme Global Change Website: www.igbp.net/news/pressreleases/pressreleases/planetarydashboardshowsgreataccelerationin-humanactivitysince1950.5.950c2fa1495db7081eb42.html

Padmakumar, K. G., Krishnan, A., Radhika, R., Manu, P., & Shiny, C. (2002). Open water fishery interventions in Kuttanad, Kerala with reference to fishery decline and ecosystem changes. *Proceedings of the National Seminar on Riverine and Reservoir Fisheries of India* (pp. 15–24). Kochi: Society of Fisheries Technologists (India).

Plunz, R. (2017). *City Riffs: Urbanism, Ecology, Place*. New York: Columbia Books on Architecture and the City, Lars Müller Publishers.

Ramsar. (2002, August 19). *Ramsar Sites Information Service*. Retrieved October 10, 2020, from https://rsis.ramsar.org/ris/1214?language=en

Redman, C. L., Grove, J. M., & Kuby, L. H. (2004, March). Integrating social science into the long-term ecological research (LTER) network: Social dimensions of ecological change and ecological dimensions of social change. *Ecosystems*, 7, 161–171.

Remani, K., Jayakumar, P., & Jalaja, T. (2010). Environmental problems and management aspects of Vembanad Kol Wetlands in South West Coast of India. *Nature Environment and Pollution Technology*, 247–254.

Robinson, J. (2006). *Ordinary Cities: Between Modernity and Development*. New York: Routledge.

Smith, A., & Stirling, A. (2010). The politics of socio-ecological resilience and sustainable socio-technical transitions. *Ecology and Society*, 15(1), 4.

The Seven Principles-Introduction. (n.d.). Retrieved November 5, 2020, from GRAID at Stolkholm Resilience Centre: https://applyingresilience.org/en/the-7-principles/

Thompson, K. (2001). *Economic and Social Issues of Biodiversity Loss in Cochin Backwaters*. Kochi: School of Industrial Fisheries, Cochin University of Science and Technology.

Turner, B. (2008). A skeptic's comments on resilience and alternative approaches to coupled human-environment systems. In *Re-framing Resilience: A Symposium Report, STEPS Working Paper 13* (pp. 9–11). Brighton: STEPS Centre.

Walker, B., Holling, C., Carpenter, S. R., & Kinzig, A. (2004). Resilience, adaptability and transformability in social – ecological systems. *Ecology and Society*. Retrieved from www.ecologyandsociety.org/vol9/iss2/art5/

6

CONCLUSION

Binti Singh and Manoj Parmar

Introduction

A recent report on climate finance by the charity WaterAid suggested that existing climate finance is not reaching the poorest and most vulnerable, who are likely to be worst affected by the climate crisis. The charity's report found that half of all countries receive less than $5 per person per year in climate finance (The Guardian, 2020). The issues of power and justice are embedded in the decision-making of resilient strategies of institutions. Resilience in cities is acknowledged both explicitly and implicitly in a number of the United Nations' 2030 Agenda for Sustainable Development (SDGs). Target 1.5 aims, by 2030, 'to build the resilience of the poor and those in vulnerable situations, and reduce their exposure and vulnerability to climate-related extreme events and other economic, social and environmental shocks and disasters' (United Nations, 2015 in Patel, 2016). Target 9.1 emphasises building resilient infrastructure while target 11 aims 'to make cities and human settlements inclusive, safe, resilient, and sustainable' and target 13.1 aims 'to strengthen resilience and adaptive capacity to climate-related hazards and natural disasters' (United Nations, 2015 in Patel, 2016) The SDGs have also extended resilience thinking to the larger questions of poverty and development goals. Recent debates on degrowth and de-globalisation also find mention in the discussion of resilient urban futures.

The story of resilience is a complex one. Cities in the Global South are increasingly becoming vulnerable with the escalating negative impacts of economic globalisation and global environmental degradation, but the resilient strategies needed to overcome these vulnerabilities do not come from outside. The problems are global; the solutions are essentially local. With escalating crises and challenges and very little outside help, the best path to stabilise our collective futures seems self-reliance, restoring faith in our people, building on our existing strengths, learning from the mistakes of others and leading by example. The COVID-19 pandemic has resulted in serious

DOI: 10.4324/9781003098461-6

shifts in global thinking on trade, supply chain and resource dependency on single countries. It is increasingly becoming clear that advanced parts of the world that were emulated for the longest time possible may actually not have all the answers. Most cities in the Global North are facing slow growth, and over the next few decades their rapidly ageing populations will steer new demands related to transport, leisure and healthcare, and existing built infrastructure will require extensive upgrading. It is best then to turn our glance to home turf and dig a little deeper for home-grown responses and solutions. Some positive news reiterates this point. A recent report pointed out that

> 'India for the first time ranks among the top ten in this year's Climate Change Performance Index (CCPI). The current levels of per capita emissions and energy use are still comparatively *low* and, along with ambitious 2030 targets, result in *high* ratings for the GHG Emissions and Energy Use categories. While the country receives an overall *medium* rating in the Renewable Energy category, India's 2030 renewable energy target is rated *very high* for its well below 2°C compatibility'.

As the report contends, the Government of India has robust policies to enable the use and expansion of renewable energy. However, the roadmap to phase out fossil fuel subsidies and reduce dependency on coal is not there yet (CCPI, 2021).

Section 2: Urban Design and Planning

The neoliberal dictates of city design and planning efforts are steered to transform the city into hard infrastructure (with attendant oil-era priorities of roads and car ownership) while people are relegated to privatised, resource-intensive and highly consumptive gated communities. This pattern of city building contradicts with our quest to support the expansion of renewable energy, reduce environment pollution and build climate resilience. Cities are all infrastructure while people relegate into customised, private spaces. This planning paradigm (based on sociological principles of structural functionalism that create segregated use of land and spaces) is now being radically reimagined. Efforts are directed to find ways and means to understand how people experience the built environment and engage with city functioning on an everyday basis. This has resulted in the re-emergence of **humanist urban thought** globally and calls for an **integrated urban science discipline**. This new imagination responds to the long-established critique of master plans, cities built from scratch, dehumanising of spaces like the office cubicle, single-use office spaces, cities layered with flyovers and expressways, business districts filled with buildings that are vertically beyond human scale, the horizontal distances of planned suburbs, peri urban fringes and the like. In the phenomenal book *Design with Nature*, author McHarg's talks of ecology as the foundation for design and planning and alludes to several examples like Big Wilds that includes large-scale conservation endeavours, such as the Yellowstone to Yukon Initiative in North America and Africa's continent-spanning Great Green Wall; Rising Tides surveys adaptation and mitigation projects that take on sea-level changes as a result of global warming, such as proposals for New York City and the North Sea; Fresh Waters addresses ensuring safe drinking water for the planet's growing population and Toxic Lands considers how to transform highly polluted sites into useful areas for people and wildlife, with examples from the Ruhr Valley in Germany and London's Olympic Park in London (Steiner, 2020).

In the 1970s, the Dutch city of Groningen was divided into four sections, and while pedestrians and cyclists were allowed to move freely, cars were prohibited from crossing between zones and were forced to take an exterior ring road that made motorised transit time-consuming – and annoying. Two-thirds of all commuting in Groningen is by bike, and this model to reduce traffic and cut down emissions has been adopted in other cities like Utrecht and Ghent. In Brussels, car access within the central Pentagon area has been restricted since the pandemic. Bart Dhondt, the Brussels municipality's alderman for mobility, explained that 60 areas across the Brussels region have been defined as potential zones for prioritised, low-emission mobility. Similarly, Amsterdam will ban diesel cars that are 15 years or older from traveling the Dutch capital's A10 ring road by the end of 2020 as part of its Clean Air Action plan and by 2030 all forms of transport in the city, including cars and motorbikes, will have to be emissions-free. Mayor Virginia Raggi of Rome has announced plans to ban diesel cars from the city centre by 2024. Madrid's zero-emissions zone already bars diesel vehicles made prior to 2006. In other metropolitan centres, city leaders address climate neutrality by proposing fundamental changes in the ways in which residents live, work and commute. Paris Mayor Anne Hidalgo unveiled a *ville du quart d'heure* – or '15-minute city' – during her 2020 re-election campaign to envision hyperlocal neighbourhood life in which every resident can find everything they need within a 15-minute walking or biking radius of their home. Smart cities specialist Carlos Moreno developed the original idea, encouraging urbanites to live in their neighbourhoods, and he explained:

> The current rhythm of life in most cities is incompatible with achieving climate neutrality by 2050. . . . In order to reduce our emissions as drastically as must be done, we need to radically transform our lifestyle. . . . If we want people to develop a sense

> of solidarity and recover a sense of integration where they live, we must let their lives play out in short distances, that allow them to develop new lifestyles based on smaller carbon footprints.

The common vision in the urban planning model of Europe's cities three decades from now involves 'lots and lots of space for walking where cars used to be parked, bikes everywhere, public transport, and maybe some shared electric cars that could even suck up surplus renewable energy while plugged into the grid' (Source: www.politico.eu/article/the-city-of-2050-less-smog-more-bikes-and-hyper-local-living/amp/?__twitter_impression=true).

Section 3: Technology, Governance and Collective Climate Wisdom

We now live in a changed world – transformed in every bit way than we had ever imagined. Material and moral densities in our times have risen in both sheer volume and complexity. Technology has played a major role in this evolution. Today, we can practically connect with anyone in any part of the world at the touch of our fingers on our phone screens. This phenomenon has created immense possibilities as well as challenges in our daily lives. As social beings, we are still trying to grapple with this altered reality. Citizens encounter complex urban challenges and contribute to mitigate or aggravate them (often inadvertently). They do, even in their most mundane daily existence and routine chores, encounter big corporations, government machinery and the like. Our lives and problems, especially those witnessed in everyday city life seem to have brought government agencies at the doorstep of citizens to the extent that one cannot function without the support and assistance of the other. These interrelationships are becoming pronounced by the day with the increasing reach of digital technologies in everyday lives. We order food with the help of apps, navigate via Google maps, consume entertainment, knowledge, news, remain fit, get services and follow trends via apps, educate and work online (especially throughout the lockdown situation during the COVID-19 pandemic). We are increasingly interfacing with the government machinery with new apps like *Arogya Setu*. Local governments and state governments have responded effectively to prevent further spread of the disease from the already severely affected dense urban areas marking the typical urban nature of the pandemic. The Integrated Command and Control Centers (ICCCs) under the Smart City Mission (SCM) have transformed into War Rooms fighting the challenges posed by the pandemic city in most part of 2020. The SCM stands on the deployment of digital technologies for dashboard governance and citizen proximity. The ICCCs work on four fronts, namely: Testing and Quarantine,

Containment, Health Advisory and Provision of Essential Services. Digital technologies have been skilfully deployed to bridge the gaps between demand and supply – farmers/food producers and housing societies, NGOs and political/relief workers and migrant labourers, medical aid providers and patients, and most importantly, frontline municipal service providers and citizens. Cities like Bhopal, Agra, Varanasi, Surat and Pune have effectively deployed digital technologies as pandemic responses through their respective ICCCs. Artificial intelligence–based video analytics have been widely used to control crowding, social distancing violations, online video consultations, one-stop shop for citizen needs and hyperlocal service delivery with tight control norms.

Digital technologies, big data, real-time data, artificial intelligence, machine learning and robotics increasingly play greater roles in real-time data, dashboard governance and the making of a Data City. Efforts and research are underway to find ways and means to make technologies and develop applications that work for the people, help in city functioning and converge to make cities more liveable. A few instances will illustrate this point. Sensor technology, particularly biosensor technology, is now cheaply available and portable. Multimer technology which originally featured in the United Nation ITU's 2015 'Measuring the Information Society Report' developed to examine crowd-sourced, quantified biometric data in a spatial context and provides experiential data about population at large spatial scale with the aim to improve spatial design through quantified, uniformly and passively collected human signals. Multitimer data include brainwave, heart rate, pedometer and GPS-collected data and analysed for a range of applications. For instance, how traffic users and city residents perceive safety and dangers, and this has been used successfully in cities like Glasgow, London, Manila, Nairobi and San Francisco. GPS-based navigation systems have become an integral part of everyday urbanism. Contemporary ECG kits and applications record brain waves and translate them into mood categories that could develop into psychogeography maps. Thus, data-driven processes open up new strategies for mapping the human experience and the ambience of places that could lead to better city functioning (Unhale & Singh, 2020).

The equalising power of technology is clearly visible despite issues of invasion of privacy and security. The question is how do we make new technologies help people develop collective climate wisdom as Carlos Moreno describes help develop new lifestyles based on smaller carbon footprints. The answer lies in the resilient strategies inbuilt in the cultural fabric of our cities (sometimes forgotten) that need to be brought back to public memory.

The foundational values of austerity, frugality, less consumption and living in harmony with natural ecologies practised in everyday living in our societies need a new affirmation from the policy circles. As Harini Nagendra draws our attention to three layers of canopy – *Polyalthia*, *Pongamia* and *Peepal* - that have created a gorgeous treescape in the neighbourhood of Koramangala and the WIPRO Electronic City urban biodiversity park in Bangalore transformed into a nature education hub with the collective efforts of experts and active citizens.

Our urban futures are indeed the handiwork of not one but several myriad stakeholders. As much as the current challenges are collective, so are the future pathways. The solutions to many of our current problems and future uncertainties are embedded in our shared ecosystems. The success of urban programmes like the *Swachh Bharat* (Clean India) Mission is attributed to coming together of citizens and governance to address crucial challenges enabled by smart technologies. The Mission was launched on 2 October 2014 for a five-year period with the objective of achieving 100% open defecation free (ODF) status and 100% solid waste management in all Urban Local Bodies (ULBs). A *Swachhata* app has been launched for complaints from citizens related to cleanliness. More than 1.55 crore citizens have downloaded the *Swachhata* app, a grievance redressal platform, and nearly, 1.79 crore complaints have been resolved (Source: http://mohua.gov.in/upload/5c987f9e0fcecUTBook25March20191compressedcompressedcompressedmin11.pdf).

The big winning cities have actually turned *Swachhata*/cleanliness mission into people's movement or *Jan Andolan* (Source: https://in.one.un.org/page/sustainable-development-goals/sdg-11/ accessed on 3 February 2020). The Economic Survey of India 2019 explains how behavioural economics and 'nudge' were used successfully in the *Swachh Bharat* Mission. For instance, the use of local '*swachhagrahis*' as foot soldiers of the *Swachh Bharat* Mission drew upon the psychology of human behaviour to nudge towards positive change. It drew attention to positive influencers, including friends or neighbours that represent role models with whom people could identify.

The Economic Survey has laid out an ambitious agenda for social and economic change using behavioural economics concepts and nudge (Source: www.thehindubusinessline.com/economy/nudging-towards-positive-change/article28286102.ece# accessed on 24 October 2019).

Mayank Midha and Megha Midha have built as many as 798 toilets across India which are being used at least 1.4 lakh times a day using the *Swachh Bharat* Mission as an opportunity to work in the urban sanitation sector deploying IOT and artificial intelligence (Source: www.thebetterindia.

com/208144/smart-toilets-garv-sanitation-self-cleaning-mayank-medha-midha/ accessed on 13 April 2020).

OMiOM Cleantech LLP www.omiom.in founded in 2016 in Mumbai develops smart and energy-efficient technologies towards Clean Water, Clean Air and Clean Energy solutions. They have developed the world's first smart hydro vehicle 'KEVAT' that aerates 60,000 litres of water per hour and captures suspended particles, microplastics and dirt. It then UV-treats the water and returns it to the lake or river.

Chakr Innovation founded in 2016 in New Delhi aims to create pioneering, sustainable and scalable technologies to combat the grave threat posed by pollution. Their product Chakr Shield can capture up to 90% of particulate matter being emitted from diesel generators and ensures that the collected particulate matter is reused as a raw material for inks and paints.

The COVID-19 pandemic has made this very evident. It has also given us an opportunity to take a step back and reflect on our growth trajectories; rethink city design, planning and governance; focus on inclusion and larger society; think collectively and in consensus and restore faith in ourselves only to emerge as more resilient and confident.

Section 4: Conclusion – the Resilience Toolkit

In this volume, we draw attention to the urbanisation trends of medium-sized cities of India and their distinct urbanisms. The sheer heterogeneity and diversity of these cities propel us to study them in detail through the lens of 'culture' – a key component of resilience and arrive at a 'typology of urban resilience'. As already noted in urban scholarship, the new wave of urbanisation currently underway in the Global South (of which India occupies a large part) unfolds under conditions of great 'uncertainty' characteristically different from the 19th- and 20th-century urbanisation in the Western/Northern Hemisphere. Most of this new wave urbanisation runs the risk of repeating the unsustainable patterns of the earlier phases after World War II, the environmental and social costs of which would be tremendous. A singular planetary urbanisation knowledge system misses out the granular realities of everyday lived experiences in cities around cultural practices and ecologies. Now is the time to dig into them to look for home-grown solutions to surmounting climate risks. We sought to contribute to the growing scholarship of southern urbanism in our limited research capacity. We also admit that resilience discourse must not be limited to theoretical constructs in academia, they have the power to be drivers of change in the way cities are imagined and administered.

As discussed in the chapters, resilience cannot be viewed as an absolute aspect at a certain scale. Rather, the implications of planning for resilience needs to take into account the various scales of application – namely City and Regional Systems, Community level and the individual level.

The concept of resilience for the Global South cities cannot be discussed without a reference to the 'five capitals' – environmental, economical, institutional, social and human. These must be read alongside the learnings from the individual cities discussed in this volume. These include social cues, cultural practices, community participation, availability of funding/resources, both immediate and long-term environmental dependencies and attendant responses unique to these cities. These responses can help enhance the resilience capacity of the city and help to identify the risk awareness, risk perceptions and the eventual action capacities for which the city administration as well as the communities need to plan. On these lines, we have attempted to develop a resilience framework that could help generate a 'Resilience Index' customised to the specificities of each rapidly urbanising geography.

We imagine this 'Resilience Index' neither as a static document nor as a canonical rule book. It must be updated regularly based on its ability of prioritising effective responses to external forces and uncertain events. In other words, for achieving its raison d'etre. The constant feedback from its application will help to reorganise the priorities and strategies through the cycle of time. The dynamic nature of operation ensures the effectiveness of this index.

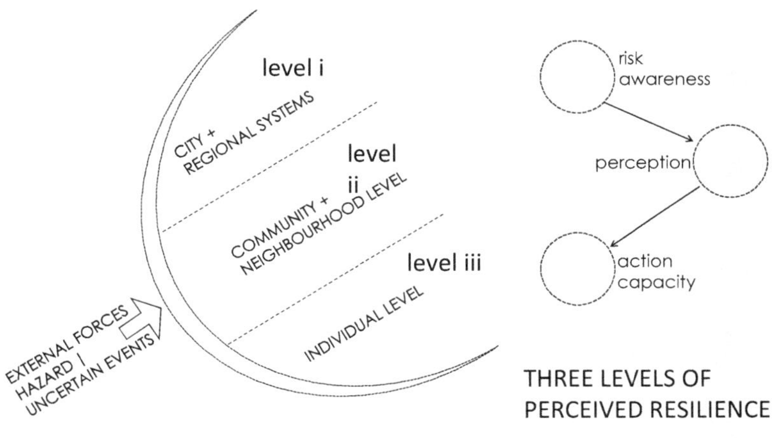

Figure 6.1 The scales of resilience planning.

Source: Authors

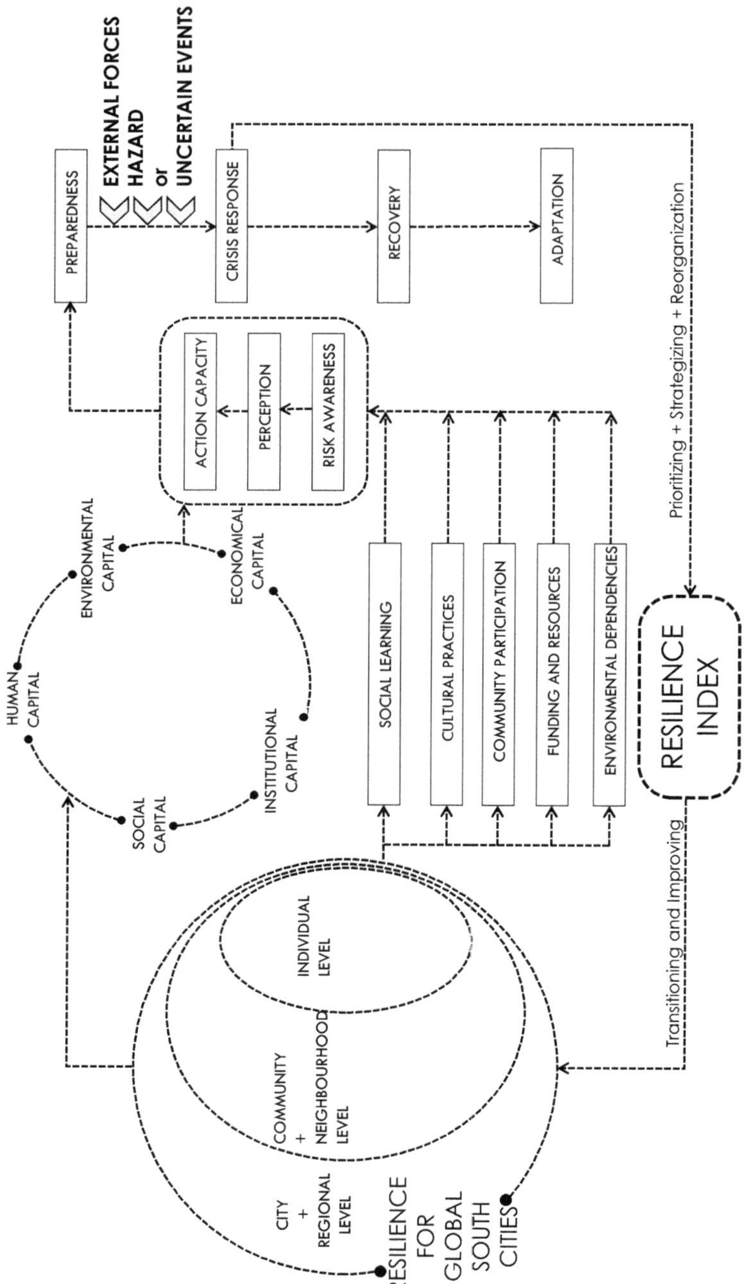

Figure 6.2 Resilience Index for the Global South city.

Source: Authors

Working on the Resilience Index may go a long way to address the uncertain futures of cities South of the Hemisphere that are resource-poor and vulnerable. Read alongside the resilience typology discussed in this volume, urban stakeholders can work their ways out towards home-grown solutions to surmounting global challenges and build better urban futures.

References

Analysis Shows Climate Finance Not Reaching Most Vulnerable. (2020, March 3). Retrieved from guardian.com: www.theguardian.com/environment/2020/mar/05/analysis-shows-climate-finance-not-reaching-most-vulnerable?CMP=share_btn_fb

Bhaduri, A. (2020, June). *Digital Tools to Tackle Water Scarcity*. Retrieved from www.indiawaterportal.org/articles/digital-tools-tackle-water-scarcity

Boffey, D. (2020, July). *Dutch city Redraws its Layout to Prepare for Global Heating Effects*. Retrieved from theguardian.com: www.theguardian.com/world/2020/jul/29/dutch-city-arnhem-redraws-layout-prepare-global-heating-effects.

Bose, S. (2016, December). *Voices from the Ground: Photovoice Research on Children's Health in the Indian Sundarbans*. Retrieved from https://dhr.gov.in/sites/default/files/FHS_SundarbansResearchBrief_Online.pd

Elliott, L. (2020, November). *UK to Make Climate Risk Reports Mandatory for Large Companies*. Retrieved from theguardian.com: https://amp-theguardian com.cdn.ampproject.org/c/s/amp.theguardian.com/environment/2020/nov/09/uk-to-make-climate-risk-reports-mandatory-for-large-companies

https://in.one.un.org/page/sustainable-development-goals/sdg-11/. (n.d.). Retrieved from https://in.one.un.org/page/sustainable-development-goals/sdg-11/

Mansata, J. (2019, November 9). *Using a Mobile App, Haqdarshak is Helping Rural India Benefit from Govt Welfare Schemes*. Retrieved from yourstory.com: https://yourstory.com/2017/11/haqdarshak-govt-welfare-schemes

Patel, R. (2016). United Nations University Centre for Policy Research, Working paper 6. *Defining the Resilient City*.

Surane, J. (2020, July). *Citi to Start Measuring How Much Carbon Emission Comes from its Loans*. Retrieved from Bloomberg.com: www.bloomberg.com/news/articles/2020-07-29/citi-to-measure-disclose-emissions-tied-to-lending-portfolio.

Technology and ODK Based Assistance during Natural Disasters. (2020, August). Retrieved from https://sdma.kerala.gov.in/wpcontent/uploads/2020/08/FloodMapping-Joy-2020.

The City of 2050: Less Smog, More Bikes and Hyper-local Living. (2020, June). Retrieved from Politico.eu: www.politico.eu/article/the-city-of-2050-less-smog-more-bikes-and-hyper-local-living/amp/?__twitter_impression=true

Unhale, S., & Singh, B. (2020). *How Will India Fix Her Urban Future?* Between Architecture and Urbanism, Mumbai.

Climate Change Performance Index (2021). Retrieved on 05-06-2021 from https://ccpi.org/country/ind/.

APPENDIX

Resilience Practices in Cities of Europe

'Citi to Start Measuring How Much Carbon Emission Comes from Its Loans'

Citigroup Inc., one of the biggest lenders to energy companies, said it will measure and disclose emissions tied to its massive lending portfolio and is working to finance $250 billion of sustainable activities by 2025 such as clean technology, water conservation or sustainable agriculture.

(Surane, Jennifer. 2020, July. www.bloomberg.com/news/articles/2020-07-29/citi-to-measure-disclose-emissions-tied-to-lending-portfolio)

'UK to Make Climate Risk Reports Mandatory for Large Companies'

UK has recently announced to make climate risk reports mandatory for large companies. Britain's first green gilt – a bond that will be floated in the financial markets during 2021 with the money raised paying for investment in carbon-reducing projects and the creation of jobs across the country.

(Elliott, Larry. 2020, Nov. https://amp-theguardian com.cdn.ampproject.org/c/s/amp.theguardian.com/environment/2020/nov/09/uk-to-make-climate-risk-reports-mandatory-for-large-companies)

'Dutch City Redraws Its Layout to Prepare for Global Heating Effects'

Under a 10-year plan, the Dutch city of Arnhem is digging up asphalt roads and creating shady areas around busy shopping districts to better prepare residents for extreme weather conditions such as downpours, droughts and intense heatwaves that make way for grasses and other plants to better dissipate heat and improve the city's absorption of rainfall.

(Boffey, Daniel. 2020, July. www.theguardian.com/world/2020/jul/29/dutch-city-arnhem-redraws-layout-prepare-global-heating-effects)

Resilience Practices in India

Story 1 – Waste to Biofuel: Experiment in Orissa

Back in early 2000s, 90% of rural villages of Orissa, especially indigenous forest communities, didn't have electric grid connection. Kinchlingi, a small village in Orissa, faced droughts and famines, water scarcity and sanitation problems due to lack of basic infrastructure like electricity. In addition, these remote villages were at the receiving end of dearth of forest or agricultural produce to sustain them. Illicit timber logging had caused large-scale deforestation, shifting agriculture practices (locally called *bogodo*) had made soils infertile and traditional crops like Niger (*Guizotia abyssinica*) were exploited for export as bird feed.

Gram Vikas, a local NGO since 1992, were on a mission of 'providing basic water supply and sanitation, infrastructure, capacity building and equitable access to secure livelihoods' to at least 1% of Orissa's population. And a Canadian NGO, CTx GREEN (community-based technologies exchange, fostering Green Energy Partnerships), partnered with Gram Vikas, to establish a biodiesel-based water pumping programme in four village communities of Orissa. The key purpose of the programme was to ensure water supply, basic sanitation services through a self-reliant energy system and simultaneously promote regeneration of the land and continued local economic opportunities within the village.

Kinchlingi being the first village of the programme, this 16-household village got biodiesel-fuelled pumpsets (3.5–5 HP) and small-scale power generation sets (2–3 kW), along with pedal-powered grinders for oilseeds (and grains), hand-operated oil presses and pedal-driven biodiesel reactors.

The village biodiesel production facility is operated by volunteers organised through the village council, supported by a bare-foot engineer trained by NGOs promoting this biomass-based energy system. Biodiesel is produced in 5-L batches, roughly one batch a week. Since May 2005, this biodiesel has been used to run a 100% biodiesel pumpset, for daily filling-up of a 9,500 L overhead tank in Kinchlingi for a total three-hour daily community pedalling. Five litres of biodiesel is required each week to run the pump 45–60 minutes a day, to provide running water to 73 Kinchlingi residents, at approximately 70 L per person per day.

Vegetable oil extracted from locally grown (and native) oil-bearing crop(s) are serving as feedstock for conversion into biodiesel. Oil cake and glycerine are valuable by-products with enhanced livelihood potential.

Village-level extraction of oil from seeds collected/cultivated locally helps return oil cake as soil nutrient to the local agro-forest ecosystem as well.

Positive Impacts (as Quoted From gnesd.org Website)

The production schedule for this very small scale technology (5-L and 20-L batch production on a bimonthly or weekly basis requiring only 20–80 kg seeds/batch, respectively) was developed in consultation with the community.

The package included organic agronomic practices to supplement local forest seeds like Karanja (*Pongamia pinnata*) and Mahua (*Madhuca indica*) with Niger (*Guizotia abyssinica*) an indigenous oilseed.

Community and private fallows belonging to other neighbouring villages were used to grow Niger consecutively, ensuring sustained agro-productivity.

Biodiesel was produced in a pedal-driven reactor that could be maintained by anyone with basic bicycle-maintenance experience. The fuel thus produced was then used in a regular pumpset, replacing diesel fuel.

Since May 2008, gravity flow water system replaced the biodiesel pumping in Kinchlingi, biodiesel is continued to be used as a back-up energy source during dry season and dependable alternate electricity (through biodiesel-fuelled battery charging) in the community.

Story 2 – Smart Village

Post-successful initiation of 99 Smart Cities under the Smart Cities Mission, much contemplation went into the issue of growing migrant population trend (from villages to cities) in India and UNDP predictions for 50% population in urban areas by 2050 (which is 19% more in four decades, 2011 – consensus division of only 31% in urban areas). In 2016, Shyama Prasad Mukherji Rurban Mission (SPMRM) was launched with the main objective to make villages 'smart' and imagine them as growth centres of the nation in line with the vision of Gandhi ji – 'The soul of India lives in its villages'.

The programme was an ambitious attempt to transform rural areas into 'Economically, Socially and Physically Sustainable spaces', or smart villages 'which would trigger overall development in the region'. Till December 2017, 266 village clusters were adopted under SPMRM programme.

Dhanora, Tehsil Bari, District Dholpur, and Rajasthan: Dhanora is a tiny village of 2,000 people, in Chambal, in the Dhaulpur District, Rajasthan. It lacked basic sanitation, internal roads and potable water. A poverty-stricken village, it also faced issues of encroachment, power fluctuation and unemployment. Till 2014, when IRS officer Dr Satyapal Singh Meena, a native of the village, along with NGO Eco Needs Foundation, later taken under SPMRM, took on development of India's first Smart village. The project started in 2016 with water conservation structures being constructed in two months. The results started showing results within three months, and after three years, the quantum of groundwater recharge is 97.49 million litres per year. Dhanora has become an 'Adarsh Gram', with a toilet and running water in every household, access to concrete roads, a sewage treatment plant for waste management and property demarcations (Source: The Better India).

And so, from 2014 to 2016, it worked on five elements for the smart village plan.

The first step was to mobilise the community.

- Retrofitting or modifying and beautifying existing structures and removing encroachments.
- Redevelopment, which involves infrastructure development.
- Greenfield, which is environment-related development.
- E-Pan which stands for electronic planning, focusing on communication and e-learning.
- Livelihood to provide learning and skills to enable people to earn.

First from Community mobilisation – *shramdaan* and fundraising by the villagers, every household built a toilet.

Story 3 – Wasteland to Forest: Story of Jharkhand

Jharkhand state derives its name from forests; however, tree felling has been a persistent problem, especially for remote areas of these forests. In an exceptional case of afforestation not by government officials or NGOs, but villagers themselves was witnessed in village Hesatu, Jharkhand's Ormanjhi administrative block, 27 km from state capital Ranchi. Near Hesatu village, lay a 365 acres wasteland that spread across 3 km radius, and villagers in 2010 internally discussed, a draft was prepared by 25–30 villagers and they decided to convert it into a forest.

By 2017, a forest of over 100,000 trees lies across this wasteland, and villagers receive annual income through our agro-forestry initiative is between Rs 40 and Rs 50 lakhs. Winning recipe was their home-grown science. They had planted trees at a distance of 8 feet from each other. Every tree has a 1.5 feet radius trench around it. Ginger and turmeric shrubs are planted around the trench, with tubers underneath. 'Three levels of plantation, three sources of income', in their own words. Inspiringly, their perspective and perseverance paid off:

Three-Tier Business Model

'Thirty percent goes to land development, thirty percent goes to the community, thirty percent to feed people who toil in the forests round the year and ten percent on welfare'. This apart, villagers have started an open school of training. 'We charge Rs 100 per day per person to learn how we work. We spend Rs 90 to feed him. Training charge is Rs 10', The latest addition to the earning is through dairy due to the forest.

> The 2016 monsoon was a boon for us. We earned around Rs 4 lakh by selling grass and bought around 70 cows. Now the villagers are earning Rs 5,000 per day by selling milk. And last year's income through forestry was around Rs 40 lakh.

The beginning was auspicious.

> We started our work six years ago in Holi. We dug the earth, ploughed it and planted vegetables the first year. With profits from vegetables, we started working on the bigger project – that of growing trees. We were already cultivating lac on kusum (*Schleichera oleosa*) and ber (*Ziziphus mauritiana*) trees on 200 acres. We added the numbers on the community forest of 365 acres.

Creating Their Own Luck

Mahavir Mahto, another farmer associated with the project, said, 'Smaller plots here and there are for vegetables. So, we are maintaining a community forest and a kitchen garden'. Alijaan Ansari, another villager, said they created their own luck. 'We have a perennial river nearby, the Domba. But, earlier, it dried up in summer. Now, the river stays full round the year'. 'Keep us away from government and NGOs', a villager laughed, not wanting to come on a quote for his candour. 'Years will pass without any good work if these two are involved. The government will waste days in passing files. NGOs can't move an inch without stakeholders meetings and documentation. We are doing well on our own', he added with a grin. But, isn't government and NGO help needed to protect valuable assets such as timber? The Hesatu team said no. 'Every tree is counted and marked as you can see. Plus, here is a canal between the forest and road that neither animals nor trucks can cross', Oraon explained.

No More Seasonal Migration by Villagers

What brings additional satisfaction to the villagers is that seasonal migration has been stopped. 'Till 2012–13, around 200 people used to leave the village to work as daily wagers in other cities. In 2014, when we realised that we have started making profits through forestry, we counselled the family members of the migrated people to stop them from going anywhere and work here instead. In the years 2015 and 2016, the rate of migration was lower. And in 2017 we can safely claim that our village is migration-free', said Sunita Devi, a villager, and active plant grower. The villagers have now identified several barren patches around Hesatu that they are now developing into small nurseries. Pahan said:

> To grow more trees, we would need more saplings and for more saplings, we would need more nurseries. We would continue to grow more trees till someone comes and says we can't grow trees in other places. But believe me, seeing the greenery around, people will appreciate rather than stopping us.

The villagers echoed something environmentalists across the globe keep reiterating. 'The cool breeze and chirping birds remind us that saving greenery is the only way to save mankind', Thakur said. The farmer then looked around in satisfaction. 'Our children too are lucky. While helping us plant trees, they learn math through measurements, colours through flowers, fruits, and vegetables. Above all, they love mother earth because each tree is like their sibling'.

Back in 2010, residents of Jharkhand's Hesatu village were sharing their worries about the wastelands in their area. During the discussion, someone proposed growing a forest of their own on the land. A draft was prepared by around 25–30 villagers and it all began.

Today, around 93 households from the village have successfully raised forest cover of more than 1 lakh trees on 365 acres of wasteland – all by themselves.

Jagnu Oraon, a village resident, told Village Square:

> We started our work six years ago on Holi. We dug the earth, ploughed it and planted vegetables the first year. With profits from vegetables, we started working on the bigger project – that of growing trees. We were already cultivating lac on kusum and ber trees on 200 acres. We added the numbers on the community forest of 365 acres.

The village of about 800 people put their skills to best use and built the forest on the wasteland and now earns an annual income of Rs 40 lakh to Rs 50 lakh through its agro-forestry initiative. Residents of the village, which comes under the Ormanjhi administrative block, around 27 km from

Ranchi, applied their knowledge and carefully planted each tree – eight feet from each other, with each tree having 1.5 feet radius trench around it.

Their income development pattern allots 30% each for land development, the community, and to feed people who work hard in the forests, the remaining 10% is utilised on welfare.

The villagers have even started an open school of training where they charge a daily fee of Rs 100 for people who want to learn how they work. Following the monsoons last year – which they say was a boon for them – the villagers even earn though dairy after earning Rs 4 lakh by selling grass and purchasing around 70 cows.

'Now, the villagers are earning Rs 5,000 per day by selling milk. And last year's income through forestry was around Rs 40 lakh', said villager Devendra Nath Thakur.

After Hesatu, the villagers have now started developing small nurseries in barren patches around the area. Thanks to their efforts, they say that even seasonal migration has come to a stop. After they started earning profits through forestry in 2014, they convinced family members of those who migrated to work on their own land instead.

'In the years 2015 and 2016, the rate of migration was lower. And in 2017 we can safely claim that our village is migration-free', said Sunita Devi, a villager (Ministry of Housing and Urban Affairs, 2020).

Story 4: *A Repository of Welfare Schemes: Story of* Haqdarshak *(Entitlements) Mobile App*

In 2016, *Haqdarshak* app was started by CEO and Co-founder Aniket Doega working primarily in Maharashtra and Rajasthan and then spreading across the entire country. The main aim for this app was to establish entrepreneurship at a grassroots level, training the underprivileged community to use the mobile platform to disseminate information about governmental schemes. The online portal as well as APP include eligibility portal which the entrepreneurs use to provide door-to-door service in registering the underprivileged eligibility and explaining to them about governmental grants. The website includes success stories of individuals gaining Indian Rupees 20,000 to 2 lakhs of grants annually supporting their living. In this way, this mobile app could aid in urban governance tackling and benefitting the huge underprivileged density in the country (Mansata, 2019).

Story 5: *Story of WASH Connect App*

Global engineering and design consultancy, Arup, and leading UK water charity, FRANK Water, have launched the WASH Connect mobile app and WASH Basins Toolkit – in supplying water, sanitation and hygiene facilities for safe, sustainable and equitable water and sanitation services. The toolkit and app use digital tools such as the KoBo Toolbox and India Space Programme technology to help farmers and professional develop accurate assessments of the water situation like underground water table, wells, tube wells and irrigation canals. These assessed conditions were mapped, and a document was submitted to state government in providing funds under various schemes. The schemes and policies under WASH are *Swach Bharat* Mission (SBM), *Swach Bharat Abhiyan* (SBA), National Rural Drinking Water Programme (NRDWP) and National Water policy of the NITI Aayog, 2018 that provide funds for livelihoods and constructing irritation, water and electricity facilities like wells and tube wells. The initial implementation of WASH project was done in states of Rajasthan, Chhattisgarh and Madhya Pradesh. The WASH Connect app aims to help farmer community to understand and manage their water better, ensuring they have access to resilient water resources for generations (Bhaduri, 2020).

How Technology Helped in Building Community Resilience in India

Story 6 – GIS- and ODK-Based Assistance During Natural Disasters: Story of Kerala

All over India, the situations of natural disaster create distress in regard to infrastructure, livestock, community and health services. In 2018 and again in 2019, Kerala faced massive floods causing damage to homes, infrastructure, farmlands and livestock. The SCMS Water Institute, Student Volunteers of SCMS Engineering College, local Panchayat body and professionals and local participants collaborated to rescue victims. The ODK (Open Data Kit) app helped professionals reaching to the individual by collecting information in the form of location, images, audio, video, signatures, barcodes, free-text multiple-choice and numeric answers. Thus, the raw information from ODK app was entered into software GIS (Geographic Information System) to generate maps highlighting inundation areas. The use of technology facilitated future uses in disaster preparedness and land use planning, for effectively using the flood plain. It also focused on increased connectivity, aiding the pace of rescue operations, supplying needs to the inaccessible victims, analysing damage, offering health kits, and donating to the cause ('Technology and ODK based assistance during natural disasters', 2020).

Story 7 – Community Resilience Built Through Photovoice Methods: Case of Sundarbans

Sundarbans known for its tigers and crocodiles is a remote inaccessible area with poor basic services like water, electricity, sanitation and health. A project Photovoice was conducted by Future Health Systems (FHS), funded by Government of UK in collaboration with the Institute of Health Management Research (IIHMR), Bangalore and local participants in capturing children's health. A technique like Photovoice is suited best to represent a mother's voice and concern, measuring her own capacity to make demands to decision-makers and take her own initiative in transforming these demands into action. All the participants were taught to hold camera and capture pictures which they thought suited best for showcasing their concerns. The major concerns like child malnutrition, scarcity of fresh drinking water, lack of access to a health provider, open defecation, pregnancy, livelihoods, poor conditions of roads and transportation systems were highlighted. A technique like photovoice can prove to be an aid in providing improved access, affordability and quality of health services for the poor through community voice ensuring maximum participation for community benefits (Bose, 2016).

References

Bhaduri, A. (2020, June). *Digital Tools to Tackle Water Scarcity.* Retrieved from www.indiawaterportal.org/articles/digital-tools-tackle-water-scarcity

Bose, S. (2016, December). *Voices from the Ground: Photovoice Research on Children's Health in the Indian Sundarbans.* Retrieved from https://dhr.gov.in/sites/default/files/FHS_SundarbansResearchBrief_Online.pd

Kalyanaraman, A. (2019, July 4). *'Nudging' towards Positive Change.* Retrieved from thehindubusinessline.com: www.thehindubusinessline.com/economy/nudging-towards-positive-change/article28286102.ece#%20accessed%20on%2024th%20Oct%202019

Mansata, J. (2019, November 9). *Using a Mobile App, Haqdarshak is Helping Rural India Benefit from Govt Welfare Schemes.* Retrieved from yourstory.com: https://yourstory.com/2017/11/haqdarshak-govt-welfare-schemes

Ministry of Housing and Urban Affairs. (2020, May). Retrieved from mohua.gov.in: http://mohua.gov.in/upload/5c987f9e0fcecUTBook25March20191compressedcompressedcompressedmin11.pdf

Steiner, F. (2020, July). *Design with Nature Now and the Pandemic.* Retrieved from www.cupblog.org/2020/07/22/design-with-nature-now-and-the-pandemic-by-frederick-steiner/?amp=1

Technology and ODK Based Assistance during Natural Disasters. (2020, August). Retrieved from https://sdma.kerala.gov.in/wpcontent/uploads/2020/08/FloodMapping-Joy-2020

INDEX

For Product Safety Concerns and Information please contact our EU representative GPSR@taylorandfrancis.com
Taylor & Francis Verlag GmbH, Kaufingerstraße 24, 80331 München, Germany

www.ingramcontent.com/pod-product-compliance
Ingram Content Group UK Ltd.
Pitfield, Milton Keynes, MK11 3LW, UK
UKHW021409220926
475569UK00022B/6
* 9 7 8 0 3 6 7 5 6 3 5 2 3 *